GLOBAL WARMING

OPPOSING VIEWPOINTS®

Other Books of Related Interest

GLOBAL WARMING

OPPOSING VIEWPOINTS®

James Haley, *Book Editor*

Daniel Leone, *Publisher*
Bonnie Szumski, *Editorial Director*
Scott Barbour, *Managing Editor*

**OPPOSING
VIEWPOINTS®
SERIES**

Greenhaven Press, Inc., San Diego, California

Cover photo: Photodisc

Library of Congress Cataloging-in-Publication Data

Global warming / James Haley, book editor.
   p. cm. — (Opposing viewpoints series)
   Includes bibliographical references and index.
   ISBN 0-7377-0909-X (lib. bdg. : alk. paper) —
ISBN 0-7377-0908-1 (pbk. : alk. paper)
   1. Global warming. 2. Global warming—Environmental
aspects. I. Haley, James, 1968–   II. Opposing viewpoints series
(Unnumbered)

QC981.8.G56 G578 2002
363.738'74—dc21
                                         2001045128
                                                 CIP

Greenhaven Press, Inc., P.O. Box 289009
San Diego, CA 92198-9009

"Congress shall make
no law...abridging the
freedom of speech, or of
the press."

First Amendment to the U.S. Constitution

The basic foundation of our democracy is the First
Amendment guarantee of freedom of expression.
The Opposing Viewpoints Series is dedicated to the
concept of this basic freedom and the idea that it is
more important to practice it than to enshrine it.

Contents

Why Consider Opposing Viewpoints?

"The only way in which a human being can make some approach to knowing the whole of a subject is by hearing what can be said about it by persons of every variety of opinion and studying all modes in which it can be looked at by every character of mind. No wise man ever acquired his wisdom in any mode but this."

John Stuart Mill

In our media-intensive culture it is not difficult to find differing opinions. Thousands of newspapers and magazines and dozens of radio and television talk shows resound with differing points of view. The difficulty lies in deciding which opinion to agree with and which "experts" seem the most credible. The more inundated we become with differing opinions and claims, the more essential it is to hone critical reading and thinking skills to evaluate these ideas. Opposing Viewpoints books address this problem directly by presenting stimulating debates that can be used to enhance and teach these skills. The varied opinions contained in each book examine many different aspects of a single issue. While examining these conveniently edited opposing views, readers can develop critical thinking skills such as the ability to compare and contrast authors' credibility, facts, argumentation styles, use of persuasive techniques, and other stylistic tools. In short, the Opposing Viewpoints Series is an ideal way to attain the higher-level thinking and reading skills so essential in a culture of diverse and contradictory opinions.

In addition to providing a tool for critical thinking, Opposing Viewpoints books challenge readers to question their own strongly held opinions and assumptions. Most people form their opinions on the basis of upbringing, peer pressure, and personal, cultural, or professional bias. By reading carefully balanced opposing views, readers must directly confront new ideas as well as the opinions of those with whom they disagree. This is not to simplistically argue that everyone who reads opposing views will—or should—change his or her opinion. Instead, the series enhances readers' understanding of their own views by encouraging confrontation with opposing ideas. Careful examination of others' views can lead to the readers' understanding of the logical inconsistencies in their own opinions, perspective on why they hold an opinion, and the consideration of the possibility that their opinion requires further evaluation.

Evaluating Other Opinions

To ensure that this type of examination occurs, Opposing Viewpoints books present all types of opinions. Prominent spokespeople on different sides of each issue as well as well-known professionals from many disciplines challenge the reader. An additional goal of the series is to provide a forum for other, less known, or even unpopular viewpoints. The opinion of an ordinary person who has had to make the decision to cut off life support from a terminally ill relative, for example, may be just as valuable and provide just as much insight as a medical ethicist's professional opinion. The editors have two additional purposes in including these less known views. One, the editors encourage readers to respect others' opinions—even when not enhanced by professional credibility. It is only by reading or listening to and objectively evaluating others' ideas that one can determine whether they are worthy of consideration. Two, the inclusion of such viewpoints encourages the important critical thinking skill of objectively evaluating an author's credentials and bias. This evaluation will illuminate an author's reasons for taking a particular stance on an issue and will aid in readers' evaluation of the author's ideas.

It is our hope that these books will give readers a deeper understanding of the issues debated and an appreciation of the complexity of even seemingly simple issues when good and honest people disagree. This awareness is particularly important in a democratic society such as ours in which people enter into public debate to determine the common good. Those with whom one disagrees should not be regarded as enemies but rather as people whose views deserve careful examination and may shed light on one's own.

Thomas Jefferson once said that "difference of opinion leads to inquiry, and inquiry to truth." Jefferson, a broadly educated man, argued that "if a nation expects to be ignorant and free . . . it expects what never was and never will be." As individuals and as a nation, it is imperative that we consider the opinions of others and examine them with skill and discernment. The Opposing Viewpoints Series is intended to help readers achieve this goal.

David L. Bender and Bruno Leone,
Founders

Greenhaven Press anthologies primarily consist of previously published material taken from a variety of sources, including periodicals, books, scholarly journals, newspapers, government documents, and position papers from private and public organizations. These original sources are often edited for length and to ensure their accessibility for a young adult audience. The anthology editors also change the original titles of these works in order to clearly present the main thesis of each viewpoint and to explicitly indicate the opinion presented in the viewpoint. These alterations are made in consideration of both the reading and comprehension levels of a young adult audience. Every effort is made to ensure that Greenhaven Press accurately reflects the original intent of the authors included in this anthology.

Introduction

"Scientists generally believe that the combustion of fossil fuels and other human activities . . . are likely to accelerate the rate of climate change."
—*United States Environmental Protection Agency, January 18, 2001*

"Most scientists do not believe human activities threaten to disrupt the Earth's climate."
—*Joseph Bast,* Heartland Policy Study, *October 30, 1998*

The global warming hypothesis originated in 1896 when Svante Arrhenius, a Swedish chemist, developed the theory that carbon dioxide emissions from the burning of fossil fuels would cause global temperatures to rise by trapping excess heat in the earth's atmosphere. Arrhenius understood that the earth's climate is heated by a process known as the *greenhouse effect*. While close to half the solar radiation reaching the earth's surface is reflected back into space, the remainder is absorbed by land masses and oceans, warming the earth's surface and atmosphere. This warming process radiates energy, most of which passes through the atmosphere and back into space. However, small concentrations of greenhouse gases like water vapor and carbon dioxide convert some of this energy to heat and either absorb it or reflect it back to the earth's surface. These heat-trapping gases work much like a greenhouse: Sunlight passes through, but a certain amount of radiated heat remains trapped.

The greenhouse effect plays an essential role in preventing the planet from entering a perpetual ice age: Remove the greenhouse gases from the atmosphere and the earth's temperature would plummet by around 60 degrees Fahrenheit (F). However, scientists who have elaborated on Arrhenius's theory of global warming are concerned that increasing concentrations of greenhouse gases in the atmosphere are causing an unprecedented rise in global temperatures, with potentially harmful consequences for the environment and human health.

In 1988, the United Nations Environment Program and the World Meteorological Organization established the Intergovernmental Panel on Climate Change (IPCC), comprising more than two thousand scientists responsible for studying global warming's potential impact on climate. According to the IPCC, the atmospheric concentration of carbon dioxide has increased by 31 percent, methane by 151 percent, and nitrous oxide by 17 percent since 1750. Over the twentieth century, the IPCC believes that global temperatures increased close to 0.5 degree Centigrade (C), the largest increase of any century during the past one thousand years. The 1990s, according to IPCC data, was the warmest decade recorded in the Northern Hemisphere since records were first taken in 1861, with 1998 the warmest year ever recorded.

Given this data, many scientists are convinced of a direct correlation between rising global temperatures and the emission of greenhouse gases stemming from human activities such as automobile use, the production of electricity from coal-fired power plants, and agricultural and deforestation practices. Concludes the IPCC in its Third Assessment Report, "The present carbon dioxide concentration has not been exceeded during the past 420,000 years and likely not during the past 20 million years. . . . In light of new evidence . . . most of the observed warming over the last 50 years is likely to have been due to the [human-induced] increase in greenhouse gas concentrations."

Based on IPCC projections that global temperatures will increase by 2.5 to 10.4 degrees F between 1990 and 2100, scientists and environmentalists are predicting that global warming will have mostly negative consequences for the world's climate. Kelly Reed of the environmental organization Greenpeace states that the "effects of global warming not only include rising global temperatures, but an increase in floods, droughts, wildfires, heat waves, intensified hurricanes and the spread of infectious disease." Accordingly, those who share Reed's view of global warming believe that the world's governments must take immediate action to limit greenhouse gas emissions.

In response to these pressures, a growing band of skeptical

scientists are questioning the validity of the global warming theory. According to these critics, the IPCC bases its predictions for rising global temperatures on faulty computer climate models, which exaggerate the climate's response to carbon dioxide and other greenhouse gases while failing to accurately reproduce the motions of the atmosphere. Explains Richard L. Lindzen, a professor of meteorology at the Massachusetts Institute of Technology, "Present models have large errors . . . [and] are unable to calculate correctly either the present average temperature of the Earth or the temperature ranges from the equator to the poles. . . . Models . . . amplify the effects of increasing carbon dioxide." Lindzen asserts that if models accurately represented the role of the major greenhouse gas—water vapor—in the climate system, they would predict a warming of no more than 1.7 degrees C if atmospheric carbon dioxide levels were doubled. This warming is significantly less than the 4 to 5 degrees C temperature increase forecasted by IPCC models under a doubling of atmospheric carbon dioxide.

Global warming skeptics also argue that natural climate fluctuation, not human activity, is responsible for the past century's rising temperatures. According to S. Fred Singer, a professor of environmental sciences at the University of Virginia, the earth's climate has never been steady and has continually warmed and cooled over the course of geologic time without any assistance from human activity. Says Singer, "The human component [in recent global warming] is thought to be quite small. . . . The climate cooled between 1940 and 1975, just as industrial activity grew rapidly after WWII. It has been difficult to reconcile this cooling with the observed increases in greenhouse gases." Singer also argues that temperature observations since 1979 are in dispute: Surface readings with thermometers show a rise of about 0.1 degree C per decade, while data from satellites and balloon-borne radiosondes [miniature transmitters] show no warming—with possible indications of a slight cooling—in the lower atmosphere between 1979 and 1997. Until the science behind the global warming theory is more settled, Singer and other skeptical scientists advocate placing no limits on the consumption of fossil fuels.

Politicians, the media, big business, scientists, and environmentalists all play conflicting roles in the global warming debate as public policy collides head-on with special interests and a complex scientific theory. *Global Warming: Opposing Viewpoints* covers the debate with a wide range of opinions in the following chapters: Does Global Warming Pose a Serious Threat? What Causes Global Warming? What Will Be the Effects of Global Warming? Should Measures Be Taken to Combat Global Warming? This anthology examines the prominent viewpoints surrounding the global warming controversy.

CHAPTER 1

Does Global Warming Pose a Serious Threat?

Chapter Preface

At the start of a heat wave in June 1988, James E. Hansen, director of the Goddard Institute for Space Studies at the National Aeronautics and Space Administration, testified before Congress that human-induced global warming as a result of greenhouse gas emissions was changing the earth's climate. His testimony focused international attention on the issue of global warming.

Since 1988 many scientists and environmentalists have sided with Hansen in the belief that global warming poses a threat to the environment. The United Nations Intergovernmental Panel on Climate Change (IPCC) projects that the earth's average surface temperature will increase by 2.5 to 10.4 degrees Fahrenheit from 1990 to 2100. Rising sea levels, extreme weather, and dislocated populations are impacts anticipated from rapid warming. Writes Dr. Kevin E. Trenberth of the National Center for Atmospheric Research, "[Global warming] is disruptive to human systems . . . because suddenly we find that weather patterns of the recent past are no longer useful guides for the future."

However, some scientists remain unconvinced by the forecasts of looming environmental catastrophe. According to critics, the computerized climate models used to predict climate change exclude important atmospheric cooling mechanisms from their data. Global temperature readings taken on the ground, from weather balloons, and from satellites contradict the projections of the models. Patrick J. Michaels, professor of environmental studies at the University of Virginia, testified before Congress in 1998 that "[James E. Hansen's] model predicted that global temperature between 1988 and 1997 would rise by 0.45 degrees Celsius. . . . [This] forecast . . . was an astounding failure, and IPCC's 1990 statement about the realistic nature of these projections was simply wrong." In Michaels's opinion, the current fallibility of computerized models raises serious doubts about the projected severity of global warming.

In the following chapter, scientists and environmentalists offer their views on whether global warming is a valid hypothesis.

> "Unintentionally . . . we have loosed a wave
> of violent and chaotic weather. . . . And the
> evidence is everywhere around us."

Global Warming Poses a Serious Threat

Ross Gelbspan

Ross Gelbspan argues in the following viewpoint that human
consumption of coal and oil and the subsequent rise of carbon
dioxide in the atmosphere have brought about unprecedented
and life-threatening global warming. He believes that the
earth is already experiencing extreme weather events as a result of this warming, such as droughts, heat waves, and wind-
storms. According to Gelbspan, immediate action must be
taken to curb further buildup of heat-trapping gases in the atmosphere through the promotion of alternative energy
sources. Gelbspan is a journalist and the author of *The Heat Is
On: The Climate Crisis, the Cover-up, the Prescription.*

As you read, consider the following questions:

1. What research finding prompted a number of scientists
   to revise upward their projections of future global
   warming, according to Gelbspan?
2. In the author's opinion, what factors have contributed to
   public skepticism concerning global warming?
3. What evidence does Gelbspan give to support his
   assertion that the buildup of carbon dioxide in the
   atmosphere contributes to global warming?

Reprinted from "Reality Check," by Ross Gelbspan, *E: The Environmental
Magazine*, vol. 11, no. 5, September/October 2000, pp. 24–26, by permission of
E: The Environmental Magazine. Subscription Dept.: PO Box 2047, Marion, OH
43306; telephone: 815-734-1242 (subscriptions are $20/year). URL:
www.emagazine.com; e-mail: info@emagazine.com.

In 1995, more than 2,000 scientists from 100 countries reported to the United Nations that our burning of oil, coal and natural gas is changing the Earth's climate. Five years later, many of the same researchers are very troubled by two things: The climate is changing much more quickly than they projected even a few years ago; and the systems of the planet are far more sensitive to even a very small degree of warming than they had realized.

Cause for Alarm

The long-anticipated federal report "Climate Change in America," was first leaked to the press in June 2000, and it forecast a dire future of disappearing alpine meadows, loss of coastal wetlands and barrier islands, and a dangerous upsurge in insect-borne diseases such as malaria. Forests will be replaced with grasslands, said the government study, and water quality problems will mount. Average U.S. temperatures, the report said, will rise by five to 10 degrees Fahrenheit (F) by the end of the 21st century.

In March 2000, researchers at the National Climatic Data Center also published alarming findings: Until the mid-1970s, the planet had been warming by one degree F per century—a rate at which most ecosystems can adapt. But for the last 20 years, Earth has instead been warming by four degrees F per century.

That same month, researchers announced that absorption of heat in the deep oceans over the last 40 years had temporarily masked the rapidly rising temperature of the planet. The findings prompted a number of scientists to revise upward their projections of future warming.

Unintentionally, we have already set in motion massive systems with huge amounts of inertia that had kept them relatively hospitable for the last 10,000 years. We have reversed the carbon cycle by about 400,000 years. We have heated the deep oceans. We have loosed a wave of violent and chaotic weather. We have altered the timing of the seasons. We are living on a very precarious outcropping of stability, and the evidence is everywhere around us.

Why, then, is there any doubt in the public mind about the reality of climate change? . . . The answer lies in the millions of dollars spent by a shrinking number of industry players to maintain the illusion of "scientific uncertainty." Also to blame is the U.S. press, which has been too lazy to look at the science and too intimidated by the fossil fuel lobby to tell the truth.

Even as villagers in Mozambique buried casualties of the horrendous rains that swamped the country in the spring of 2000, ExxonMobil declared in an ad on the op-ed page of *The New York Times:* "Some . . . claim that humans are causing global warming, and they point to storms or floods to say that dangerous impacts are already under way. Yet scientists remain unable to confirm either contention." But that is categorically untrue.

The Greening Earth Society, a creation of the Western Fuels Coal Association, takes a slightly different tack. Citing the opinion of a few "greenhouse skeptics"—most of whom are on its payroll—Western Fuels trumpets the idea that more warming and more carbon dioxide (CO_2) is good for us because it will promote plant growth and create a greener, healthier natural world.

They forget to mention that peer-reviewed science indicates the opposite. While enhanced CO_2 creates an initial growth spurt in many trees and plants, their growth subsequently flattens and their food and nutrition value plummets. As enhanced carbon dioxide stresses plant metabolisms, they become more prone to disease, insect attacks and fires.

Media "Balance"

The media, however, continue to report the issue as though the science was still in question, giving the same weight to the "greenhouse skeptics" as they do to mainstream scientists—all in the name of "journalistic balance." Real balance, reflecting the weight of opinion within the scientific community, would accord mainstream scientists about 85 percent of an article and leave a couple of paragraphs to the skeptics. Only recently have journalists begun to dismiss the industry-sponsored naysayers.

Nevertheless, the news media still find it very difficult to cover the biggest story of the century and, perhaps, in modern history, thoroughly and consistently. Asked about this failure, a ranking editor at one network replied, "We did include a line like that once. But we were inundated by calls from the oil lobby warning our top executives that it is scientifically inaccurate to link any one particular storm with global warming." The editor concluded, "Basically, our executives were intimidated by the fossil fuel lobby."

And resistance to the solution is staggering. We need to be generating as much energy from non-carbon sources by the year 2050 as we generate from coal, oil and natural gas today, according to a peer-reviewed article in the journal *Nature*. That means, say the authors, that we need to begin to move toward a global energy transition within this decade and we need to pursue it "with the urgency of the Manhattan Project," which developed the atomic bomb in less than three years.

A Simple and Inexorable Process

While climate science can be dizzyingly complex, the underlying facts are simple. Carbon dioxide in the atmosphere traps heat. For the last 10,000 years, we enjoyed a constant level of CO_2—about 280 parts per million (ppm)—until about 100 years ago, when we began to burn more coal and oil. That 280 has already risen to 360 ppm—a concentration that has not been seen for 400,000 years. It is projected to double to 560 ppm later in this new century, correlating with an increase in the average global temperature of three to seven degrees F. (For perspective, the last Ice Age was only five to nine degrees colder than the current climate.)

Evidence for the build-up of heat-trapping carbon dioxide abounds: The 11 hottest years on record have occurred since 1983; the five hottest consecutive years were 1991 to 1995; 1998 was the hottest year on record; the decade of the 1990s was the hottest at least in this past millennium; and the planet is heating more rapidly than at any time in the last 10,000 years. On this point the science is unambiguous: to allow the climate to re-stabilize requires worldwide emissions reductions of 70 percent.

The politics are almost as unambiguous. In December 1999, Great Britain's chief meteorologist and the head of National Oceanic and Atmospheric Administration (NOAA) declared that the climate situation is now "critical," urging the world to begin now to reduce its use of carbon fuels. The issue of climate change is the subject of serious debate only in the United States. When 160 nations met in Kyoto, Japan, in 1997 to forge a climate treaty, not one government took issue with the science.

Characteristics of Several Greenhouse Gases

	Carbon Dioxide	Methane	Chloro-fluorocarbon	Nitrous Oxide
Preindustrial concentration	280 ppm*	0.8 ppm	0	0.288 ppm
Current concentration	360 ppm	1.7 ppm	500 ppt**	0.310 ppm
Accumulation rate (annual)	0.5%	0.9%	4%	0.25%
Residence time	3 years	10 years	100 years	150 years

*parts per million   **parts per thousand

John Horel and Jack Greisler, *Global Environmental Change: An Atmospheric Perspective*. New York: John Wiley & Sons, Inc., 1997.

Since then, findings which link the warming to our burning of coal and oil have become so robust that a number of countries are moving toward solutions regardless of what happens in the U.S. The Dutch, for one, are creating a plan to reduce emissions by 80 percent over the next 40 years. Germany is contemplating 50 percent cuts in the future. Britain announced it will cut emissions by 21 percent below 1990 levels in the next 12 years.

The view of the world's business leaders is moving on the same trajectory. A vote by executives of the world's largest corporations, finance ministers and heads of state who attended the World Economic Forum in Davos, Switzerland in February 2000 was remarkable. When conference organizers polled participants on which of five different trends were most troubling, the participants overrode the choices and declared climate change to be by far the most threatening issue facing humanity.

Some of the world's largest oil and auto companies also acknowledge the perils of climate change and are positioning themselves for a new non-carbon economy. John Browne, CEO of British Petroleum-Amoco, announced his company is preparing to do $1 billion a year in solar commerce by the decade's end. Shell has created a new core company to produce renewable energy technologies. Ford and Daimler-Chrysler, together with Ballard Power Corporation, have entered a $1 billion joint venture to produce fuel-cell-powered cars in the next three years. And both Honda and Toyota are marketing 60- to 70-mile-per-gallon climate-friendly hybrid cars in the U.S.

A Question of Liability

The strongest corporate concerns about climatic instability come from the world's property insurers. During the 1980s, the insurance industry lost an average of $2 billion a year to damages from droughts, floods, storm surges, sea level rise and other extreme weather events. In the 1990s, it lost an average of $12 billion a year—$89 billion in 1998 alone. "Man-made climate change will . . . bring us increasingly extreme natural events and consequently increasingly large catastrophe losses," an official of Munich Reinsurance said recently.

While die-hard elements of the fossil fuel lobby continue to attack the findings of mainstream science, they are becoming increasingly isolated. For years, the Washington, D.C.–based Global Climate Coalition (GCC) waged a campaign against mainstream science. But its corporate membership has hemorrhaged. The GCC has been abandoned by Ford, Daimler-Chrysler, General Motors, the Southern Company and Texaco.

The very few independent scientists who still question whether global warming is caused by human activity focus on discrepancies between temperatures in the upper levels of the atmosphere and on the ground. That doubt was put to rest in January 2000 when a panel of the National Academies of Science reported that such differences "in no way invalidates the conclusion that the Earth's temperature is rising."

But the case for climate change rests on a far broader base than computer models and atmospheric dynamics alone. Add the unceasing bombardment of extreme weather events wreaking havoc all over the world.

Take, for example, 1998, which began with a January ice storm that left four million people without power in Quebec and northern New England. For the first time, rainforests in Brazil and Mexico actually caught fire. The summer brought killer heat waves in the Middle East, India and Texas, where residents suffered through a record 29 consecutive triple-digit days. Mexico experienced its worst drought in 70 years.

1999 saw a record-setting drought in the Mid-Atlantic states, with declarations of disaster in six. A heatwave in the Midwest and northeastern U.S. claimed 271 lives. Hurricane Floyd visited more than $1 billion in damages on North Carolina. A super-cyclone in eastern India killed 10,000 people. That winter, mudslides and rains in Venezuela claimed 15,000 lives. Unprecedented December windstorms swept northern Europe, causing more than $4 billion in damages. And Boston experienced a record 304 consecutive days with no snow.

Conditions are shifting rapidly, meteorologically and otherwise. Most of the public is now intuitively aware of climate change—and extremely worried about changes in the weather. Growing numbers of corporate leaders are realizing that the remedy—a world-wide transition to renewable and high-efficiency energy sources—would, in fact, create a huge surge of jobs and a dramatic expansion in the total wealth of the global economy. And national as well as grassroots political activists are at last making the climate crisis the focus of campaigns. It is too slow and too small—but it is a beginning. The issue is not whether we will mobilize around the climate crisis, but whether we will do it in time.

"[Global-warming] promoters have yet to present convincing evidence that global warming poses any environmental threat."

Global Warming Does Not Pose a Serious Threat

S. Fred Singer

In the following viewpoint, S. Fred Singer argues that contrary to dire scenarios of floods and famine, global warming remains scientifically unproven and does not pose a threat to the environment and human welfare. Singer maintains that politicians, the media, environmentalists, and bad science have joined forces to foist a global warming scare on the American public. According to Singer, these factions promote a fear of global warming in order to secure research grant money and justify the expansion of government control over personal behavior. Singer is professor emeritus of environmental sciences at the University of Virginia and president of the Science and Environmental Policy Project in Fairfax, Virginia.

As you read, consider the following questions:
1. In what manner do the media fail to fairly present the global warming debate, according to Singer?
2. Why is there a lack of scientific consensus on the causes of global warming, according to Singer?
3. In Singer's opinion, how do government funding agencies influence the global warming debate?

Reprinted, with permission, from "Cool Planet, Hot Politics," by S. Fred Singer, *American Outlook*, Summer 2000, pp. 38–40.

To the average American, the greatest threat facing the United States in the twenty-first century is probably something on the order of a new Cold War with China as the chief opponent, nuclear missiles launched by rogue nations, Islamic fundamentalism, terrorists releasing appalling new biological weapons, or cyber-warfare against the nation's banks, air-traffic control systems, and other economic targets. But to the Clinton/Gore White House, the greatest threat was something far different. As former Secretary of State Warren Christopher assured his audience at Stanford University in May 1996, the main threat is climate change produced by the burning of fuels that keep us warm, light our homes, and run our cars. Tens of thousands of global-warming promoters jet around the world annually to attend UN conferences in exotic locales while preaching the gospel of "renewable energy" based on solar and wind power, both of which are currently impracticable and unlikely to be usable for many years. But after logging thousands of miles and burning millions of gallons of fuel, these promoters have yet to present convincing evidence that global warming poses any environmental threat.

The theory of global warming, which is actually a century old, is that increased levels of greenhouse gases in the earth's atmosphere cause net increases in global temperatures. Though the doomsday scenarios generated by proponents of this theory have not been verified by climatologists, the political community of Washington, D.C., has made the alleged phenomenon a national priority; and the UN has made it a global one. The chief U.S. protagonist is certainly former Vice President Al Gore, the author of *Earth in the Balance*, which became the bible of environmental extremism. . . . Gore's disciples were everywhere in the Clinton administration, chiefly in Carol Browner's Environmental Protection Agency (EPA) but also in the Departments of State and Defense.

Political Jostling

Congress appears to be squarely opposed to Gore's radical brand of environmentalism. In July 1997, the U.S. Senate voted 95-0 for the Byrd-Hagel Resolution, which opposes any global-warming mitigation scheme that would damage the U.S. economy or let other nations off the hook. The

Clinton White House promptly reinterpreted this bipartisan rejection of mandatory cutbacks in the use of fuels as allowing it to agree to "meaningful reductions" by "key nations," with the definitions to be supplied later. In December 1997, at the Kyoto conference of the Parties to the UN Climate Treaty, the U.S. delegation ignored the Senate resolutions altogether and accepted a 7 percent reduction of fossil fuel use, which works out to a whopping 35 percent cut by the year 2010. A year later, a minor State Department official quietly signed the Kyoto Protocol on behalf of the United States, though it was not submitted to the Senate for ratification.

The Clinton administration and Congress played a cat-and-mouse game, each trying to win the policy war and finally sway public opinion. EPA made a feeble attempt—scotched by Congress—to label carbon dioxide, the chief target of the Kyoto Protocol, as a "pollutant," which would have permitted the federal government to control all emissions of it, under the Clean Air Act. Despite that setback, the Clinton White House managed to bleed money out of its $2-billion-a-year climate-research program to study the possible effects of a putative global warming on different geographical areas and population groups of the United States. It held eighteen regional workshops, each exhibiting some regional suitable horror scenario such as sea-level rise in Florida, disappearing glaciers in Montana, and floods, droughts, and tornadoes in the Midwest. In reaction, Rep. Joe Knollenberg (R-MI) has added amendments to six appropriation bills, to prohibit the use of funds for implementing, in any way, the purposes of the Kyoto Protocol. He claims strong bipartisan support, including that of heavyweights such as Rep. John Dingell (D-MI).

Business and Labor

Both business and labor are internally divided on the issue. Energy companies tend to be against Kyoto, but foreign-based oil firms such as British Petroleum and Shell are at least speaking favorably about going along, even as they continue to sell petroleum products and search for more crude oil. In beggar-thy-neighbor fashion, natural-gas firms and pipeline companies hope to gain an advantage. This attitude can be found also in the ads of the nuclear industry, which is eager to burnish its public image as a non-emitter of greenhouse gases. The business opponents of Kyoto are organized around the Global Climate Coalition, while the Kyoto supporters have the International Climate Change Partnership. The Clinton White House tried to split industrial opposition further by offering "early credits" to firms that reduce emissions voluntarily. These credits could become valuable if Kyoto is ever ratified and includes a trading program in emission rights that can be sold to firms that find it difficult to reduce emissions. Obviously, companies that have earned early credits are more likely to become political boosters for Kyoto.

Reprinted by permission of Chuck Asay and Creators Syndicate, Inc.

Support for Kyoto from labor is divided along white-collar/blue-collar lines. The United Mine Workers have already come out strongly against Kyoto because the agreement heavily penalizes coal use. Other unions, however, envision a migration of manufacturing jobs out of the country if the United States accepts Kyoto's restrictions on energy use. White-collar workers, teachers, and public employees, of course, do not have this problem and can afford to cater to their partisan interests, which are mostly Democratic. Many academics and media people tend to have a similar mindset, and the possible loss of millions of blue-collar jobs leaves them unconcerned.

Environmental Propaganda

Environmental pressure groups, however, provide most of the real excitement. Greenpeace is no longer just focusing on seal pups and dolphins, and Ozone Action has stopped worrying about the stratospheric ozone layer. Global warming is the new cash cow for these organizations, and everyone wants a piece of it. The newest kid on the block, in fact, is really no kid at all but a well-funded propaganda operation to promote the Kyoto Protocol to industry, set up by the Pew Charitable Trusts, the beneficiary of ultraconservative Joe Pew's Sun Oil money. Of course, there are think tanks on the other side as well (such as the Cato Institute and the Competitive Enterprise Institute), spreading the message that the best information available from climate science contradicts the alleged need for drastic policies certain to cause great economic harm. Needless to say, these groups don't get any government money.

Media Bandwagon

The media play a crucial role in the ongoing debate. While they overwhelmingly believe that global warming is a looming problem that calls for fairly drastic measures, most try to mention, somewhere, an opposing viewpoint, but they often identify it with industry and therefore, by implication, as self-serving and not worth listening to. The *Washington Post* and *New York Times* have had environmental reporters who practiced blatant advocacy rather than journalism. These extremists have been replaced by less obvious partisans.

An interesting example is provided by *Brill's Content*, which prides itself on being above the fray. It investigated a case in which the *International Herald Tribune (IHT)* published an op-ed signed by two scientists, George Woodwell and John Holdren, featuring ad hominem attacks on scientists skeptical toward the global warming hypothesis. The op-ed, it turned out, had been submitted by Ozone Action (OA) and may even have been written by that organization—in which case the *IHT* should certainly have informed its readers of the authors' interest, as *Brill's* noted. After the editor of *IHT* complained, *Brill's* ombudsman Bill Kovach investigated. He was assured by the scientists that they had written the piece, but he found that OA had "helped with research." Unfortunately, his report leaves open the question of exactly who drafted the op-ed. If the "research" consisted of several well-chosen paragraphs that the scientists signed, perhaps after minor changes, they did not "write" the op-ed in the accepted sense of the term. Kovach, however, failed to follow up on that question and dismissed as mere "coincidences" *eight* correspondences between the op-ed and a propaganda flyer put out by OA during the same week. One example of these correspondences: Both op-ed and flyer attacked the skeptics' position in the same words, claiming that it "dissolves under close scrutiny."

Scientific societies and journals tend to express a range of views, depending on who's in charge. For example, when Dr. Jane Lubchenco, an Oregon State marine ecologist, served a term as president of the American Association for the Advancement of Science, she used her office to promote White House positions on global warming, cooperating with Ozone Action, until journalists got wise to her doings. Scientific journals often change their policies depending on who the editor happens to be. *Chemical & Engineering News*, the flagship publication of the American Chemical Society, permits its environmental reporters to practice advocacy rather than journalism. Science organizations need to practice adult supervision to prevent such abuses.

Scientists Skeptical

Although the mass media have come to a consensus on global warming, the scientific community has not. Surface temperature data do show a warming since about 1850, the end of the "Little Ice Age," but most of it occurred before 1940, after which the climate cooled for more than three decades. Weather satellite data, the only truly global measurements, show no current warming, in direct disagreement with the best computer-created climate model predictions. Critics can say "garbage in, garbage out" regarding the computer predictions, but climate models are the only tools available for predicting future climate conditions. Unless validated by scientific observations, the models cannot justify drastic actions that will inevitably lead to economic decline, to solve a "problem" that has not been observed in reality but only predicted by computers fed information by fallible human beings with their own judgments of what is important to consider. No wise person would buy an expensive insurance policy without some evidence of risk. Moreover, renowned economists assure us that a warming of the planet would actually bring benefits, not losses. Agriculture, for example, can only benefit from more rain with fewer severe storms, milder winters, longer growing seasons, and higher levels of carbon dioxide. And contrary to the conventional wisdom, global warming would not speed up the rise of sea level but might actually slow it down because increased evaporation from the oceans leads to more precipitation and increased ice accumulation in the polar regions.

The lack of scientific consensus on the causes and possible effects of global warming is easily demonstrated. Many scientists show "concern" in public but voice doubts in private. Government funding agencies, which support much scientific research, are unlikely to support a proposal unless it expresses deep concern about global warming and explains how the study will save the world. Other scientists don't have such constraints. The "dwindling band of skeptics" who consider climate warming the "empirical equivalent of the Easter Bunny" (as Al Gore put it) is growing rapidly. In fact, a real thorn in Gore's side is the "Leipzig Declaration," which grew out of a scientific conference held in that city in November 1995 and has been signed by more than a hundred climate scientists. After highlighting the shaky science supporting the global warming scare and the absence of any scientific consensus, it concluded: "In a world in which poverty is the greatest social pollutant, any restriction on energy use that inhibits economic growth should be viewed with caution. For this reason, we consider 'carbon taxes' and other drastic control policies—lacking credible support from the underlying science—to be ill-advised, premature, wrought with economic danger, and likely to be counter-productive." A group of broadcast meteorologists and a number of state climatologists have signed similar documents. Even more impressive is the 1998 "Oregon Petition" against the Kyoto Protocol, which was signed by nearly 20,000 scientists.

Partisan Politics and Global Warming

A puzzling fact about the whole global-warming debate is that although it should be a question of science, proponents of the Kyoto Protocol tend to be Left-wing, liberal, and Democratic, while opponents are overwhelmingly Right-wing, conservative, and Republican. Perhaps the best explanation for this pattern is that the policies proposed to mitigate global warming would certainly lead to massive expansion of government control of economic operations and personal behavior, affecting the use of energy, automobiles, and everything from recycling to enforced conservation. It would also require a much greater involvement of the United Nations, overriding many aspects of national sovereignty. The Left considers these measures good and indeed necessary because they subscribe to Al Gore's vision of the world. A majority of the Right, however, strongly opposes anything that impinges on personal freedom.

The public, strangely enough, has not shown much interest in global warming. In a 1989 survey of American adults, only 35 percent expressed concern. By 1997, the response was even lower; only 24 percent said that they worried a "great deal" about global warming. Citizens have strong preferences for personal cars, SUVs, and private transportation, and they hate it when the electricity goes out. They apparently consider energy use absolutely essential to modern daily life. The public also has widely varying views about the proper role of government, which were tested during the November 2000 presidential elections. [President] George W. Bush, although ambivalent about the climate science, opposes Kyoto. Bush supporters were hoping that Gore would make the climate issue the centerpiece of his campaign. Here, oddly enough, they were on the same wavelength as the environmentalists. During an Earth Day event, Greenpeace activists heckled the then-Vice President with hoots of "Al Gore, read your book!" *Earth in the Balance* may finally get the close scrutiny it deserves.

"The planet and the nation are certain to experience more than a century of climate change, due to . . . greenhouse gases already in the atmosphere and the momentum of the climate system."

The Magnitude of Global Warming May Become Extreme

National Assessment Synthesis Team

The following viewpoint describes the projected impact of global warming on the United States. It was excerpted from a report prepared by the National Assessment Synthesis Team of the U.S. Global Change Research Program. According to the report, referred to as the Assessment, the United States is expected to undergo rapid and potentially extreme climate change during the twenty-first century if no actions are taken to reduce world greenhouse gas emissions. The Assessment details how rising temperatures will likely strain the U.S. populace with environmental, social, and economic disruptions. Global warming may also unravel political and social stability abroad, posing a challenge for global security and U.S. foreign policy, according to the authors.

As you read, consider the following questions:

1. According to the Assessment, temperatures in the United States will rise by how many degrees in the next 100 years?
2. What possible unanticipated physical and biological impacts of global warming are mentioned by the Assessment?
3. How might the construction of seawalls harm the environment, according to the Assessment?

Reprinted from *Climate Change Impacts on the United States: The Potential Consequences of Climate Variability and Change*, by the National Assessment Synthesis Team, U.S. Global Change Research Program, Washington, DC, 2000.

Long-term observations confirm that our climate is now changing at a rapid rate. Over the 20th century, the average annual US temperature has risen by almost 1°F (0.6°C) and precipitation has increased nationally by 5 to 10%, mostly due to increases in heavy downpours. These trends are most apparent over the past few decades. The science indicates that the warming in the 21st century will be significantly larger than in the 20th century. Scenarios examined in this Assessment, which assume no major interventions to reduce continued growth of world greenhouse gas emissions, indicate that temperatures in the US will rise by about 5–9°F (3–5°C) on average in the next 100 years, which is more than the projected *global* increase. This rise is very likely to be associated with more extreme precipitation and faster evaporation of water, leading to greater frequency of both very wet and very dry conditions.

Global Warming's Harmful Effects

This Assessment reveals a number of national-level impacts of climate variability and change including impacts to natural ecosystems and water resources. Natural ecosystems appear to be the most vulnerable to the harmful effects of climate change, as there is often little that can be done to help them adapt to the projected speed and amount of change. Some ecosystems that are already constrained by climate, such as alpine meadows in the Rocky Mountains, are likely to face extreme stress, and disappear entirely in some places. It is likely that other more widespread ecosystems will also be vulnerable to climate change. One of the climate scenarios used in this Assessment suggests the potential for the forests of the Southeast to break up into a mosaic of forests, savannas, and grasslands. Climate scenarios suggest likely changes in the species composition of the Northeast forests, including the loss of sugar maples. Major alterations to natural ecosystems due to climate change could possibly have negative consequences for our economy, which depends in part on the sustained bounty of our nation's lands, waters, and native plant and animal communities.

A unique contribution of this first US Assessment is that it combines national-scale analysis with an examination of the potential impacts of climate change on different regions of the US. For example, sea-level rise will very likely cause further loss of coastal wetlands (ecosystems that provide vital nurseries and habitats for many fish species) and put coastal communities at greater risk of storm surges, especially in the Southeast. Reduction in snowpack will very likely alter the timing and amount of water supplies, potentially exacerbating water shortages and conflicts, particularly throughout the western US. The melting of glaciers in the high-elevation West and in Alaska represents the loss or diminishment of unique national treasures of the American landscape. Large increases in the heat index (which combines temperature and humidity) and increases in the frequency of heat waves are very likely. These changes will, at minimum, increase discomfort, particularly in cities. It is very probable that continued thawing of permafrost and melting of sea ice in Alaska will further damage forests, buildings, roads, and coastlines, and harm subsistence livelihoods. In various parts of the nation, cold-weather recreation such as skiing will very likely be reduced, and air conditioning usage will very likely increase.

Potential Benefits vs. Unanticipated Impacts

Highly managed ecosystems appear more robust, and some potential benefits have been identified. Crop and forest productivity is likely to increase in some areas for the next few decades due to increased carbon dioxide in the atmosphere and an extended growing season. It is possible that some US food exports could increase, depending on impacts in other food-growing regions around the world. It is also possible that a rise in crop production in fertile areas could cause prices to fall, benefiting consumers. Other benefits that are possible include extended seasons for construction and warm weather recreation, reduced heating requirements, and reduced cold-weather mortality.

Climate variability and change will interact with other environmental stresses and socioeconomic changes. Air and water pollution, habitat fragmentation, wetland loss, coastal erosion, and reductions in fisheries are likely to be compounded by climate-related stresses. An aging populace nationally, and rapidly growing populations in cities, coastal areas, and across the South and West are social factors that interact with and alter sensitivity to climate variability and change.

1000 Years of Global Carbon Dioxide (CO_2) and Temperature Change

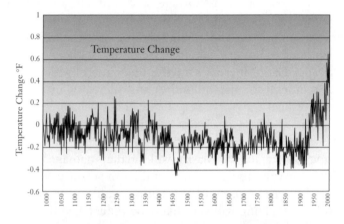

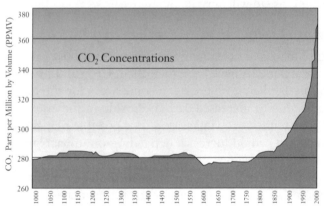

National Assessment Synthesis Team, *Climate Change Impacts on the United States: Potential Consequences of Climate Variability and Change.* Washington, DC: U.S. Global Change Research Program, 2000.

There are also very likely to be unanticipated impacts of climate change during the next century. Such "surprises" may stem from unforeseen changes in the physical climate system, such as major alterations in ocean circulation, cloud distribution, or storms; and unpredicted biological consequences of these physical climate changes, such as massive dislocations of species or pest outbreaks. In addition, unexpected social or economic change, including major shifts in wealth, technology, or political priorities, could affect our ability to respond to climate change.

Adaptation Will Become Necessary

Greenhouse gas emissions lower than those assumed in this Assessment would result in reduced impacts. The signatory nations of the Framework Convention on Climate Change are negotiating the path they will ultimately take. Even with such reductions, however, the planet and the nation are certain to experience more than a century of climate change, due to the long lifetimes of greenhouse gases already in the atmosphere and the momentum of the climate system. Adapting to a changed climate is consequently a necessary component of our response strategy.

Adaptation measures can, in many cases, reduce the magnitude of harmful impacts, or take advantage of beneficial impacts. For example, in agriculture, many farmers will probably be able to alter cropping and management practices. Roads, bridges, buildings, and other long-lived infrastructure can be designed taking projected climate change into account. Adaptations, however, can involve trade-offs, and do involve costs. For example, the benefits of building sea walls to prevent sea-level rise from disrupting human coastal communities will need to be weighed against the economic and ecological costs of seawall construction. The ecological costs could be high as seawalls prevent the inland shifting of coastal wetlands in response to sea-level rise, resulting in the loss of vital fish and bird habitat and other wetland functions, such as protecting shorelines from damage due to storm surges. Protecting against any increased risk of water-borne and insect-borne diseases will require diligent maintenance of our public health system. Many adaptations, notably those that seek to reduce other environmental stresses such as pollution and habitat fragmentation, will have beneficial effects beyond those related to climate change.

Global Challenges to US Security

Vulnerability in the US is linked to the fates of other nations, and we cannot evaluate national consequences due to climate variability and change without also considering the consequences of changes elsewhere in the world. . . . For example, conflicts or mass migrations resulting from resource limits, health, and environmental stresses in more vulnerable nations could possibly pose challenges for global security and US policy. Effects of climate variability and change on US agriculture will depend critically on changes in agricultural productivity elsewhere, which can shift international patterns of food supply and demand. Climate-induced changes in water resources available for power generation, transportation, cities, and agriculture are likely to raise potentially delicate diplomatic issues with both Canada and Mexico.

This Assessment has identified many remaining uncertainties that limit our ability to fully understand the spectrum of potential consequences of climate change for our nation. To address these uncertainties, additional research is needed to improve understanding of ecological and social processes that are sensitive to climate, application of climate scenarios and reconstructions of past climates to impacts studies, and assessment strategies and methods. Results from these research efforts will inform future assessments that will continue the process of building our understanding of humanity's impacts on climate, and climate's impacts on us.

*"Fools in Washington are rushing in . . . to
tackle a problem [global warming] that
science cannot confirm."*

There Is No Evidence That the Magnitude of Global Warming Will Be Extreme

James K. Glassman

James K. Glassman argues in the following viewpoint that there is no credible evidence that global warming is imminent. He takes issue with a report on climate change (referred to as the "Assessment") prepared by the National Assessment Synthesis Team of the U.S. Global Change Research Program. Glassman considers the Assessment's gloomy predictions of global warming's impact on the United States to be further evidence of the political gamesmanship surrounding the global warming debate. According to Glassman, the Assessment is intended to bolster government efforts to raise taxes while reducing emissions of carbon dioxide. Glassman is a trustee of the Reason Foundation, a libertarian public policy research organization.

As you read, consider the following questions:
1. According to Glassman, how did the timing of a draft release of the Assessment make its climate predictions appear politically motivated?
2. What do many scientists find troubling about the Assessment, according to Glassman?
3. What are two contradictory findings of the Canadian and Hadley climate models, in Glassman's opinion?

Reprinted from "Global Climate Scare: Fools Rush In," by James K. Glassman, *Reason Online*, October 9, 2000, by permission of the author.

In the days ahead, you will be hearing more and more about something called the National Assessment Synthesis Report. It is an analysis of the effects of global climate change on the environment, agriculture, water, health, society, biological diversity, and on and on.

The Politics of "Gloom and Doom"

The report [Assessment] was prepared by the U.S. Global Change Research Program, a group that includes representatives from different federal agencies. An advance draft was rushed into print—in time for [former] President Clinton to talk about it in his State of the Union Address. But the comments on the draft by scientific reviewers were so devastating that the report was held up. Now, it's back—just in time for the November 2000 election. The idea is that a global warming scare would benefit [former Vice President] Al Gore's candidacy, and the report provides just the kind of gloom and doom to do the trick.

A revised draft was made public for comment in June 2000, and the gist of it is reflected in headlines like this one in the *Washington Post:* "Drastic Climate Changes Forecast: Global Warming Likely to Cause Droughts, Coastal Erosion in U.S., Report Says." Peter Jennings said on ABC TV: "The draft of a report to Congress says that global warming over the next century is going to be severe enough in many parts of the country to end winter as we know it."

Such scary tales will make it easier for Gore and other politicians to convince Americans to embrace their agenda of heavy taxes and onerous controls on businesses and individuals in an effort to reduce emissions of carbon dioxide—even though there is no evidence that global warming is a serious threat, is imminent, or has manmade causes.

The report is so deficient that the chairman of the House Science Committee asked the [Clinton] White House to hold it up. No dice. Then, on Oct. 5, 2000, the Competitive Enterprise Institute, along with Reps. Joseph Knollenberg (R-Mich) and Jo Ann Emerson (R-MO) and Sen. James Inhofe (R-OK) filed a lawsuit contending that the report violated several federal laws, including one requiring open meetings. The suit asked that the Assessment not be made public until it is fixed.

The problem with the report is that it's bad science. Its general conclusions are certainly frightening. In the words of an article in *Science* magazine, greenhouse gasses will cause the U.S. to "warm, affecting everything from the western snowpacks that supply California with water to New England's fall foliage." In addition, wrote reporter John Fialka of the *Wall Street Journal*, "Rising sea levels and heavy rains will force coastal cities to spend billions of dollars to redesign subways, tunnels, dams and sewage-treatment systems." And, wrote Fialka, higher temperatures will "trigger the spread of waterborne diseases and red tides [and] a northward expansion of malaria, dengue fever and other diseases."

But the report lacks the detail to back up many of its contentions, due to what reporter Richard A. Kerr of *Science* called "the rudimentary state of regional climate science." And many scientists are outraged by the distortions in the report, among them Mike Slimak and Joel Scheraga of the Environmental Protection Agency itself. The two wrote the health report in the Assessment but complain that "many statements have a rather extremist/alarmist tone and do not appear to fairly reflect the scientific literature, the historical record, or the output of extant models." John Christy, a respected scientist at the University of Alabama said of the report, "I saw no attempt at scientific objectivity. This document is an evangelistic statement about a coming apocalypse, not a scientific statement." And, said Dr. Jae Edmonds of the Batelle National Laboratory, "The current version of the report reads more like an advertising supplement to *Time* magazine than a National Assessment." Meanwhile, six scientists from the Joint Program on the Science and Policy of Global Change at the Massachusetts Institute of Technology stated that, among other things, the "Assessment did not conduct a rigorous and scientific assessment of uncertainty. The inferences in the text that lead the reader to believe that such an analysis was undertaken should be eliminated."

Predictions Based on Inconclusive Models

The problem is that report sounds as if the dire results it predicts are much more likely than the state of current science can possibly say they are. For example, there are two widely used climate models, one from the Canadian Climate Center and the other from the U.K. Hadley Center for Climate Research and Prediction. On many key points, the two models produce very different conclusions.

An Ideal Global Climate

Strangely enough, if you believe the prognosticators right now, we live in the "best of all possible worlds," at least as far as climate goes. In the 1970s, many scientists worried about global cooling. The Department of Transportation organized a multiyear research effort involving hundreds of scientists and economists to evaluate its effects. The researchers found that a cooling of the world would reduce living standards. Since many of those same forecasters now predict doom from warming, we are obviously living on the edge between a world that is too hot and one that is too cold. Given that mankind, over the last million or so years, has evolved in climates that were both hotter and colder than today's, how is it that we in the 20th century are so fortunate as to have been born into the ideal global climate?

Thomas Gale Moore, *Climate of Fear: Why We Shouldn't Worry About Global Warming*. Washington, DC: Cato Institute, 1998.

The Canadian model, for instance, predicts that precipitation in the Northeast will fall by 10 percent, but the Hadley model says it will rise by 24 percent. The Canadian model says that conifer forests and grasslands in the West will increase, while the Hadley model says they will decrease. In the Southeast, the Canadian model says that crop yields will fall, but the Hadley model says they will rise.

No wonder the EPA's own website warns, "Complicated models are able to simulate many features of the climate, but they are still not accurate enough to provide reliable forecasts of how the climate may change; and the several models often yield contradictory results."

But that didn't stop [former] President Clinton in June 2000 from saying that the report "suggests that change in climate could mean more extreme weather, more floods, more droughts, disrupted water supplies, loss of species, dangerously rising sea levels."

He added, with characteristic hyperbole and sentimentality, "This is about science, this is about evidence, this is about the things that are bigger than all of us and very much about our obligation to these children here to give them a future on this planet."

But the National Assessment is not about science and about evidence. It is about politics and about laying the groundwork for taking drastic steps that will almost certainly raise energy costs and put downward pressure on the standard of living in the United States—not to mention in developing countries.

For decades, bad science has been placed in the service of political ends. But nothing like this. Fools in Washington are rushing in—with drastic solutions—to tackle a problem that science cannot confirm, let alone accurately assess.

> *"The GCM [general circulation model] modelers . . . have long endeavored to subvert actual knowledge of the Earth's climate with bogus predictions of catastrophic, man-made global warming."*

The Theory of Global Warming Is Not Scientifically Credible

Gene Barth

Gene Barth is a senior research technologist at the University of Chicago Medical Center. In the following viewpoint, Barth contends that the theory of global warming is unsupported by solid scientific data. According to Barth, the scenarios of environmental devastation due to man-made global warming are based entirely on general circulation model (GCM) climate simulations. In Barth's opinion, the climate simulations distort the mechanisms that determine global temperatures and therefore produce results that contradict more objective evidence showing that no global warming is occurring.

As you read, consider the following questions:

1. What is a negative consequence of the incorrect climate predictions supplied by the GCMs, according to Barth?
2. According to Thomas Moore as cited by the author, how much higher were CO_2 concentrations in the atmosphere sixty million years ago?
3. What is a major mechanism by which the earth is cooled that the GCMs exclude, in Barth's opinion?

Reprinted, with permission, from "The Distorted World of Climate Models," by Gene Barth, *The Intellectual Activist*, February 3, 1998, pp. 3–10.

At the United Kingdom Meteorological Office (UKMO) in Bracknell, England—illuminated by florescent light, bathed in a flow of refrigerated air, and electrified on an industrial power grid—sits one of the most sophisticated computing machines devised by the mind of man. Ironically, this machine is a tool in an attempt to extinguish the lights, halt the flow of refrigerated air, and black out the power grids of the world by constricting man's combustion of vital fossil fuels.

The machine at Bracknell is a supercomputer, attended by a bevy of PhDs and graduate students. Its prodigious computing capacity is dedicated to around-the-clock execution of a giant simulation of the Earth's atmosphere called a general circulation model, or GCM. The GCM's attendants claim that the output of the simulation is a true, detailed account of the Earth's future climate, on a time scale of decades.

In reality, the GCM modelers at UKMO—and a handful of sites in the US—have long endeavored to subvert actual knowledge of the Earth's climate with bogus predictions of catastrophic, man-made global warming.

The consequences of this subversion are potentially grave. The GCM climate simulations are the *sole* basis for the claim by the United Nations' Intergovernmental Panel on Climate Change (IPCC) that solid science, accepted by a worldwide consensus of scientists, backs the specter of an Earth parched (or flooded) by global warming. In turn, [former] Vice President Al Gore and a horde of other environmentalists at December 1997's "climate summit" in Kyoto, Japan, cited only the authority of the IPCC in calling for a reduction in US combustion of fossil fuels to 7% below 1990 levels by the year 2012. At current growth rates, that means slashing this country's use of fossil fuels by fully one third of projected 2012 levels.

Since the combustion of fossil fuels constitutes man's primary means of generating power—from 1300-megawatt coal-fired electrical generating units, to automobiles, to leaf blowers—the GCM climate simulations constitute the sole scientific rationalization for the environmentalists' attempt to impose a crippling, worldwide scarcity of energy as a chronic, permanent feature of man's life on earth. A glimpse of this future was not long in coming. Less than a week after the close of the Kyoto gathering, Norway indefinitely suspended construction of two gas-fired power plants. Norwegian Environment Minister Guro Fjellanger declared that there was no room for new power plants after the Kyoto summit.

How has this disastrous myth been given a veneer of scientific validity?

Objective Measurements Show No Warming

The global warming dogma certainly gains no support from measurements of the Earth's climate. Indeed, the more objectively the temperature of the Earth is sampled, the more sharply the resulting temperature record contradicts claims of rapid global warming. Tom Karl of the National Oceanic and Atmospheric Administration has carefully culled historical temperature records for the continental United States. He scrupulously eliminated from these records those stations around which towns grew over the course of the century. Towns and cities are warmer than countryside; thus, when a temperature station originally in the country becomes part of a town, it inaccurately reports a higher temperature for the entire region it represents. Karl's corrected record, the Historical Climate Network, shows no warming during the last 60 years—yet the bulk of man-made carbon dioxide (CO_2) has been released into the atmosphere since 1950.

Nor is any support for global warming dogma forthcoming from the heavens. According to physicist S. Fred Singer, satellite data, unbiased by urban warming and available since 1979, "have shown no climate warming—contrary to all expectations. According to computer models cited by [the IPCC], a substantial warming of nearly one degree Fahrenheit should have occurred during this 18-year period." Further, a "completely independent set of measurements [made by] instruments launched around the world by weather balloons show exactly the same temperature trends as the satellite instruments."

Finally, the Earth's past is also cold comfort for global warming dogma. Thomas Moore, a senior fellow at the Hoover Institution, sums up the CO_2 record: "Carbon dioxide concentrations may have been up to sixteen times higher about sixty million years ago without producing runaway greenhouse effects. Other periods experienced two to four times current levels of CO_2 with some warming. Scientists have been unable to determine whether the warming preceded or followed the rises in carbon dioxide." Seconding Moore's observations, research physicist Sherwood Idso writes, "Much of the world was a degree or two warmer 6,000 to 1,000 years ago . . . when the CO_2 content of the atmosphere was fully 80 ppm [parts per million] less than it is today."

Climate Models: "The Data Don't Matter"

The real evidence concerning the Earth's temperature and its relation to CO_2 levels overwhelmingly contradicts the claims of catastrophic global warming. How, then, can the environmentalists ignore this data? The rationalization they need is provided by the GCMs. Like medieval monks engaging in arguments over how many angels can dance on the head of a pin, the environmentalists immerse themselves in the study of computerized climate models, which they substitute for observation of reality.

For example, environmentalist Stephen Schneider, then at the National Center for Atmospheric Research, the sponsoring agency for one of five major GCMs, declared contemptuously: "Looking at every bump and wiggle of the record is a waste of time. . . . So, I don't set very much store in looking at the direct evidence." Back at the UKMO, Christopher Folland sets even less store in the evidence. At a 1991 meeting of climatologists, Folland was shown a temperature time line showing that the northern hemisphere warmed only 0.9°F from 1880 through 1950, then stopped warming altogether. Next to the temperature graph, a second time line showed the 250% increase in annual fossil fuel carbon release from 1950 through 1980, the period in which there was no warming. [See graph.] To this direct refutation of the GCMs' prediction that man's CO_2 emissions are baking the planet, Folland commented simply, "The data don't matter." Fresh from his work on the IPCC's first executive summary (released in 1992), Folland explained that "we're not basing our recommendations upon the data; we're basing them on the climate models."

The climate models manage to controvert the historical

facts about the climate because they are deliberately constructed to distort the real mechanisms that determine global temperatures. The real atmosphere of Earth acts in part as a layer of insulation, absorbing infrared radiation from the sun-warmed surface and then reemitting that radiation to further warm the surface. The constituents of the atmosphere that absorb and reemit infrared radiation, and thereby insulate the Earth, are called "greenhouse gases." But water vapor is far and away the most important greenhouse gas, accounting for 98% of the insulating effect of the atmosphere; carbon dioxide and other minor greenhouse gases account for only the remaining 2%. Thus, a doubling of CO_2 from present levels, by itself, would increase the infrared flux at the Earth's surface by only 2.5 watts per square meter—about what a flashlight would deliver illuminating the same patch of ground. By comparison, the solar flux at mid-day in summer is 1000 watts per square meter, 400 times greater.

Positive Feedback Distorts Results

It is the job of the GCM models to amplify the wisp of additional power from a doubling of atmospheric CO_2 into a planet-baking threat. The models accomplish this by inventing two enormous "positive feedback" mechanisms.

In the world according to the GCMs, the relative humidity (RH) of air never changes. (This premise is called the "fixed RH profile.") But the absolute capacity of air to hold water vapor does increase with temperature, doubling every 20°F. Thus, to maintain the same relative humidity—and thereby satisfy the arbitrary assumption of a fixed RH profile—any increase in air temperature in the GCM world must instantly conjure up an increase in the absolute amount of water vapor. As a greenhouse gas, this additional water vapor warms the GCM world further, resulting in still more water vapor, and so on.

The second positive feedback built into the GCMs is the assumption that the temperature of the Earth's atmosphere decreases with height at a constant rate. (This premise is called the "fixed tropospheric lapse rate.") To maintain this arbitrary assumption, any increase in the temperature of air at the Earth's surface must instantly conjure up a corresponding temperature increase throughout the entire vertical column of the atmosphere above it. And of course, given the fixed RH profile, the now warmer upper atmosphere instantly conjures up more water vapor, which, as a greenhouse gas, further warms the Earth—and so on, in an upward spiral. Working together, the fixed RH profile and the fixed tropospheric lapse rate constitute a positive water vapor feedback that amplifies what would otherwise be a 1.8°F warming due to a doubling of atmospheric CO_2 into a sizzling 7.2°F warming. In the world of the GCMs, the tail, CO_2, wags the dog, water vapor.

Ignoring the Role of Convection

Meteorologist Hugh Ellsaeser points out that these assumptions turn the real facts about the atmosphere on their head by systematically excluding the enormous cooling power of atmospheric convection. Convection is the rise of warmer, lighter air from the surface of the Earth into the atmosphere. Far from being a mechanism for conjuring up more water vapor and thus more warming—as in the positive water vapor feedback of the GCMs—convection acts to dry and cool pockets of hot, damp air.

This cooling is achieved in three related ways. First, warm air ascends into the upper regions of the atmosphere, above most of the heat-trapping water vapor, where it can efficiently radiate its heat into space. Second, as the rising air expands and cools, water vapor condenses; the now drier air traps less heat from the lower regions of the atmosphere. Finally, the condensation of water vapor produces cloud cover that reflects solar radiation back into space before it can warm the ground. Together these effects are powerful. "Any objective analysis of tropical convection," Ellsaesser writes, "leads to the conclusion that it is a highly effective . . . planetary temperature regulator," i.e., a *negative* water vapor feedback.

Indeed, MIT climatologist Richard Lindzen observes that convection is the major mechanism by which the Earth is cooled. In the absence of convection, the Earth's atmosphere would be stagnant and would be cooled at all levels only by the radiation of heat into space. The temperature of such an

Earth would be far more sensitive to the concentration of greenhouse gases. Greenhouse gases earn their name by trapping infrared radiation—that is, by impeding radiative cooling. Thus, on an Earth cooled only by radiation, any increase in greenhouse gases necessarily causes a hotter climate, since the Earth must radiate its heat through more "insulation."

Carbon Dioxide Emissions Compared with Temperature Relative to 59°F in the Northern Hemisphere

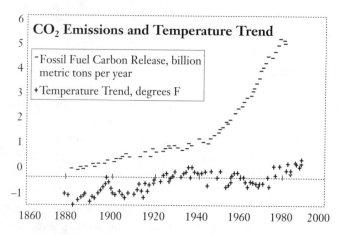

Gene Barth, *Intellectual Activist*, February 1998.

Just how much hotter would the Earth be in the absence of convection? At present levels of greenhouse gases, the Earth would burn up at 171°F if radiation were its only mechanism of cooling. Thus, the action of convection "short circuits" the greenhouse effect by a whopping 112°F, cooling the Earth to a livable average temperature of 59°F.

Given these facts, the basic practice of the GCM modelers has been to ignore and suppress the overwhelming role of convection, biasing their models toward radiative cooling. The fictitious Earth they create is hotter and more sensitive to the level of greenhouse gases—essential properties if such a model is to be a tool of environmentalist dogma and therefore "policy relevant" and deserving of federal cash. . . .

In 1975 Syukuro Manabe and Richard Wetherald published the result of a GCM climate simulation that incorporated this fantastically unreal positive water vapor feedback. The simulation obediently popped out a whopping global warming of 7.2°F for a doubling of atmospheric CO_2. . . . Attracted by an apocalyptic forecast tricked out as sophisticated science, the environmentalist policy makers fixed on the Manabe-Wetherald GCM as the "scientific centerpiece" of a global-warming campaign. . . .

Reaching a "Consensus," Minus the Critics

But fraud on such a massive scale could not have been successful without the sanction of genuine scientists. Thus, in 1988 the United Nations Environment Program and the World Meteorological Organization (another organ of the UN) spawned the Intergovernmental Panel on Climate Change. The most important function of the IPCC has been to co-opt honest scientists, and even vocal and able critics of the global warming dogma, by embracing them as "contributors" or "reviewers" of its reports. The IPCC leaders then write these scientists—along with their doubts and objections—out of the report's executive summary, while still counting them as part of the "consensus" for apocalyptic global warming. Environmentalists around the world have then pointed to this manufactured consensus, declared the global warming dogma "settled science," and, in a growing crescendo, called for a world-wide regulatory stranglehold on man's combustion of fossil fuels.

But the very dependence of the global warming dogma on an illusion of scientific respectability is the key to its potential downfall. To their lasting credit, a small group of scientists have refused to go along with the bogus "consensus." They have not remained content to enumerate the scientific errors of the GCMs in academic journals and then sheepishly assume their places on the IPCC's contributor list. Instead, they have sought to thoroughly debunk the global warming dogma, and to expose and combat the methods of the environmentalists.

Cooling the Dogma with Better Science

This principled defense of science has begun to cool the global warming hysteria on two fronts. First, realistic models of convection—demonstrated over a decade ago in the academic literature—are pushing the sham Manabe-Wetherald positive water vapor feedback out of the GCMs. Consequently, the GCMs have dramatically cooled, dropping from predicted warming as high as 9.4°F to warmings as low as 1.4°F—and in one model 0°F, i.e., no enhanced warming from additional CO_2. When the "scientific centerpiece" of the global warming dogma is removed, its constellation of harmful consequences—melting polar ice caps, the spread of tropical diseases, etc.—evaporate back into the thin air from which they came.

In addition, 18 years of high-quality satellite data has made its way into the scientific literature, where it sternly contradicts all expectations based on the GCMs. Exults climatologist Patrick Michaels, "What was heresy four years ago—that the climate models were forecasting way too much warming—is now consensus. Scientists who said satellites were finding no warming whatsoever were ridiculed, and journalists were remonstrated by Mr. Gore to ignore them. Now they're being lionized because they saw what was happening first."

Finally, and most important, scientists are refusing to be herded into the IPCC "consensus." In 1992, and again in 1995, scores of distinguished climate scientists and meteorologists signed declarations of dissent from the IPCC "consensus." In fact, every poll ever taken of scientists who study the atmosphere has shown scant support for apocalyptic global warming. A number of scientists have summoned the courage to publicly declare the IPCC process corrupt. . . .

Unfortunately, the US Senate—which will now be charged with accepting or rejecting the Kyoto agreement—may not be listening. . . .

If the US Senate leadership is to develop any moral outrage, any courage, any impulse to proudly spurn the UN's demand that man give up fire, it will depend on the rational conviction that the science is settled—settled *against* the myth of apocalyptic global warming. If the fraud at the root

of the GCMs can be stripped of its veneer of scientific respectability, the demand for massive restrictions on fossil fuels will be seen for what it is: a naked call for human sacrifice, a sacrifice, in the words of one defender of science [S.
Fred Singer], in the name of "climate 'disasters' that exist
only on computer printouts and in the feverish imagination
of professional environmental zealots."

VIEWPOINT

6

"Union of Concerned Scientists (UCS) is convinced . . . that the earth's temperature is rising."

The Theory of Global Warming Is Scientifically Credible

Union of Concerned Scientists

The Union of Concerned Scientists (UCS) is dedicated to a worldwide reduction in the emission of heat-trapping greenhouse gases. In the following viewpoint, the UCS maintains that there is a scientific consensus supporting the theory that the global climate is warming due to the increasing atmospheric concentration of CO_2 and other gases resulting from human activities. The UCS believes that the climate models forecasting global warming have been closely scrutinized and have the support of prestigious scientific panels. The impacts of global warming remain uncertain, but serious environmental consequences may arise if emissions reductions are not undertaken, in the opinion of the UCS.

As you read, consider the following questions:
1. Why is the global warming debate so contentious, according to the UCS?
2. How does the UCS address concerns that the scientific consensus on global warming is politically motivated?
3. What does the Intergovernmental Panel on Climate Change conclude about human activities and global warming?
4. In the opinion of the UCS, what is the best way to improve public understanding of global warming issues?

Reprinted, with permission, from "The Science of Global Warming," a fact sheet from the Union of Concerned Scientists' website at www.ucsusa.org/environment/gw/science.html.

The science of global warming, or climate change, is currently more hotly contested in the media and among policymakers than the science of any other environmental issue. Even though the vast majority of scientists agree on the basics of climate science, the public portrayal of the issue has left many people with the impression that scientists are so deeply divided that remedial action is premature. Moreover, some anti-environmentalists have used global warming as a symbolic issue to suggest that scientists have such divergent and confused views of global trends that it is pointless to look to science for guidance.

Union of Concerned Scientists (UCS) is convinced—along with over 2500 of the world's leading climate scientists, economists, and risk analysis experts—that the earth's temperature is rising and that its climate has changed over the last century. We also believe, as do these scientists, that the rise in temperature and change in climate are being caused in part by human activities. To understand the debate, one must understand the ways scientists reach consensus.

Consensus Versus Certainty

A fundamental reason for the contentious debate is that climate change discussions necessarily focus on the future, something that by definition cannot be known with certainty. The extremely complex nature of the earth's climate system makes predictions about future climate patterns especially tricky and open to debate. But this does not mean that scientists are confused about global warming. For one thing, it is quite clear that certain gases, such as carbon dioxide and methane, play a crucial role in determining the earth's climate by preventing heat from escaping the atmosphere. Researchers have also been able to document that the increased concentration of such gases in the atmosphere results from human activities such as the burning of fossil fuels, deforestation and land degradation, cattle ranching, and rice farming.

What scientists cannot know for sure is the exact impact on the earth's climate of these rising gas levels. However, there has been striking agreement among most scientists about what is most likely to occur. Computer models developed by climatologists suggest that the planet will warm up over the next generations, representing a more rapid climate change than at any time in recorded history. The current best estimate is that a doubling of carbon dioxide concentrations from preindustrial levels will cause temperatures to rise between 1.8 and 6.3 degrees Fahrenheit (1.0–3.5°C).

Although the exact consequences of this warming necessarily remain uncertain, climatologists have predicted certain impacts of global warming. Scientists predict that sea levels will rise, making coastal groundwater saltier, endangering wetlands, and inundating valuable land and coastal communities. In many places, changes in precipitation patterns could have a greater impact than rising temperatures, with some regions likely to receive additional precipitation and other inland arid and semiarid regions likely to suffer from decreased rainfall. As wild plants and animals experienced changed climate conditions, some of them would be unable to adapt or migrate to new locations. There could be serious human health impacts as well, both from heat stress and from the spread of disease-carrying insects currently confined to tropical regions.

These and other impacts of global warming mean that different countries and regions would have very different experiences. In some higher latitudes, agricultural productivity could actually rise even as farming was devastated in other areas. Small island states and countries such as Bangladesh, with extensive low-lying coastal areas, are especially vulnerable to sea-level rise. Overall, the United Nations Environment Programme concludes that "poor countries may be the biggest losers of all," since they may "lack the necessary resources for protecting themselves." In general, global warming would place additional stress on a global environment that is being seriously harmed by various human activities.

The climatologists' models and research contributions from workers in other fields have undergone vigorous review through the peer review process and have been closely examined and supported by prestigious panels appointed by such organizations as the National Academy of Sciences. In order to evaluate the huge amount of published results, the World Meteorological Organization and the United Nations Environment Programme established, in 1988, the Intergovernmental Panel on Climate Change (IPCC) to assess the latest scientific and technical information about global warming.

The Role of the Intergovernmental Panel on Climate Change

The task of the Intergovernmental Panel is to assess the scientific and technical information about climate change in a comprehensive, transparent, and objective manner. The reports of the Panel are made possible through the cooperation of the scientific community around the world. Hundreds of scientific and technical experts were involved in preparing the Panel's 1995 report, and literally thousands more were engaged to provide objective peer review. The participants were drawn from academia, from private and national research laboratories, and from nongovernmental organizations. The Panel makes a concerted effort to include the broadest possible range of scientific opinion. Indeed, the entire credibility of the Panel in the eyes of both governments and the scientific community rests on its commitment to providing contemporary, balanced scientific information that truly reflects the state of human understanding of climate change science.

Still, the IPCC has often been accused of being an implacable monolith and of having imposed a dogma of contrived consensus for politically motivated reasons. Some scientists, even within the climate community, have expressed reservations regarding the "consensus science" produced by the Panel. Though the Panel has no apparatus to refute these claims and worries, one may address these concerns by considering the process by which the reports are produced. The Panel has no permanent bureaucracy except a small Secretariat, which is responsible for logistical coordination. The Panel relies entirely on the support of the scientific and technical communities to produce its reports. Peer review is an essential component of the assessment process. To ensure integrity, participation was considerably expanded for the *1995 Second Assessment* to over 2,500 scientists representing more than 80 countries, up from 200 scientists representing 40 countries for the original report. Among those reviewing the *Second Assessment Report* were scientists who generally dissent from the evidence of global climate change. The views presented in the report stem from the analysis of over 20,000 articles from the relevant literature. The scientists tried to reconcile competing views through peer review if possible. When they could not reach a consensus, the scientists characterized the disagreements and identified the issues that needed clarification through additional research. Thus, the *Second Assessment Report* should be seen for what it is: a massive, policy-neutral review of the current state of understanding of climate change science.

The Need for Research and Clarification

Of course, important questions still remain. For example, scientists are intensively investigating the interactions of water vapor, clouds, oceans, and sea ice in the global climate system in order to refine their computer models. In addition, scientists are continuing to work to pin down the likely impact of global warming on plant ecosystems, since higher carbon dioxide levels, higher temperatures, and changed precipitation patterns would have complicated effects on particular cultivated crops and forest communities. In general, scientists need to know more about how the climate would change region by region and what the regional and local consequences of climate change would be. Virtually all scientists agree that more research is needed to answer such questions and to improve scientific understanding of global warming. As research progresses, current climate projections will continue to be refined.

Has the Climate Already Changed?

Historical records show that the climate has changed in the last one hundred years. Among the trends established were the following:

- Global mean surface temperature has increased 0.5 to 1.1°F (0.3–0.6°C) with night-time minimums increasing more than day-time maximums.
- The 10 warmest years since 1860 have all been since 1980.
- Sea level has risen between 3.9 and 10 in. (10–25 cm) because of thermal expansion of the oceans.

- The mean annual air temperatures of Antarctica have increased steadily since the 1950s.
- The persistent 1990 to mid-1995 warm phase of the El Niño–Southern Oscillation event was exceptional in the 120-year record of the phenomenon.

In considering the indirect historical record, the Intergovernmental Panel concluded that the 20th century was at least as warm as any period since 1400, while a recent study found the mean temperature of 1901–1990 is higher than any 90-year interval since AD 914.

Thus, the evidence of a statistically unusual change in climate has recently become more compelling. But the more important issue the Intergovernmental Panel has to address is whether this change can be attributed to human activities. If there is a signal of human interference, it is superimposed on a background noise of natural climate variation. The results indicate that the observed warming is unlikely to be entirely natural in origin. On the basis of these latest results, the Panel, in the *Second Assessment Report*, makes an unprecedented, though qualified, attribution of the observed climate change to human causes. Though the human signal is still building and somewhat masked within natural variation, and while there are key uncertainties to be resolved, the Panel concludes that "the balance of evidence suggests that there is a discernible human influence on global climate."

The Intergovernmental Panel developed a series of six future climate change projections. They were based on a range of estimations of heat-trapping-gas and aerosol emissions and assumptions about future population, energy use, economic growth, land use changes, and so on. The recent advances in the understanding of climate science led to some slight revisions of the scenarios in the *Second Assessment*, but has increased the Panel's confidence in the use of global circulation models to project future climate change. The elaboration of six scenarios is intended to convey the range of possibilities while accounting for the important uncertainties still outstanding. Predicting the future remains inherently risky.

The revised Panel forecasts of the increased global mean temperature by the year 2100 range from 1.8 to 6.3°F

About Heat-Trapping-Gas Emissions

Human activities produce emissions of several gases that scientists believe will contribute to global warming. The chart below lists emissions levels for these gases, along with the human-caused emission sources. The last column of the chart lists the "direct global warming potential" of each gas—a measure the Intergovernmental Panel has devised to show "the possible warming effect on the surface-troposphere system arising from the emission of each gas relative to carbon dioxide." The chart shows, for example, that the Panel concludes that each ton of methane will have 11 times the global warming impact over a hundred-year period as a ton of carbon dioxide. Even though total emissions of chlorofluorocarbons are quite small compared with emissions of carbon dioxide, their impact is significant since their global warming potential is so large. Nevertheless, carbon dioxide emissions still account for about half the total global warming potential of emissions from human sources.

	Anthropogenic sources	Total global emissions, 1991 (in metric tons)	Direct global warming potential over 100 years
Carbon dioxide	Burning of fossil fuels, cement manufacture, deforestation and other land-use changes	26,073,000,000	1
Methane	Livestock, wet rice agriculture, solid waste, coal mining, oil and gas production	250,000	11
Nitrous oxide	Nylon production, nitric acid production, biomass burning, cultivated soils, automobiles with three-way catalysts	Accurate data not available	270
Chlorofluoro-carbons (CFCs)	Chemical products and processes, including refrigeration, industrial solvents, blown-foam insulation	400	3400–7100
CFC substitutes (HCFCs and HFCs)	Under development as substitutes for CFCs	Minimal	1200–1600

J.T. Houghton et al., *Climate Change 1992: The Supplementary Report to the IPCC Scientific Assessment* (New York: Cambridge University Press, 1992), 14–15; *World Resources Institute, World Resources 1994–95* (New York: Oxford University Press, 1994), 362–66.

(1–3.5°C), with a "best estimate" scenario projecting 3.6°F (2°C). The latter estimate is about one-third that forecast in 1990, primarily because of lower emission estimates for carbon dioxide and chlorofluorocarbons and the effect of aerosols. In all scenarios, the projected warming would probably be greater than any similar event in the last 10,000 years. The warming over the next 100 years is expected to raise the average sea level by a best estimate of 19.7 in. (50 cm), with a range of 6 to 37.5 in. (15–95 cm). Sea level would continue to rise at comparable rates in centuries beyond 2100 as the earth-atmosphere system continued to equilibrate. The thermal inertia of the oceans ensures that sea levels would rise even if heat-trapping gases and global mean temperatures were stabilized by 2100.

Warmer global temperatures are also expected to produce a more vigorous hydrological cycle, with the inherent prospect of more severe drought and/or floods in geographical areas prone to those types of events. Some sources project an increase in precipitation intensity, that is, more frequent rain- and snowfall of extreme magnitudes. Additional future adverse impacts from projected global warming include the possibilities of coastal flooding; severe stress on forests, wetlands and other ecosystems as plant species' ranges are altered on an accelerated time scale; dislocation of agriculture and commerce for similar reasons; and damage to human health from changes in the dissemination of serious infectious diseases.

The Science/Policy Intersection

Climate scientists—and others in related fields—will continue their research to refine our understanding of the earth's complex atmospheric system. However, despite the advances they have made, or perhaps because of them, climate science itself is under increasing attack in the media and policy forums. Ultimately, the best way to improve public understanding of global warming issues and to create a more receptive atmosphere for policy action is for scientists to repeatedly, patiently, and strategically present accurate, credible information to the media and policymakers. UCS's Sound Science Initiative is doing just that.

Part of what scientists must do is to explain why it is unrealistic and unnecessary to expect total scientific certainty before taking appropriate action to address the threat of global warming. As climatologist Stephen Schneider notes, "I'm not 99 percent sure, but I am 90 percent sure [that the climate is changing]. Why do we need 99 percent certainty when nothing else is that certain? If there were only a 5 percent chance the chef slipped some poison in your dessert, would you eat it?" The only time we will be 100 percent certain that scientists' beliefs about climate change are accurate is when we observe dramatic changes to the climate. But then it will be too late to take the most effective remedial measures.

Periodical Bibliography

The following articles have been selected to supplement the diverse views presented in this chapter. Addresses are provided for periodicals not indexed in the *Readers' Guide to Periodical Literature*, the *Alternative Press Index*, the *Social Sciences Index*, or the *Index to Legal Periodicals and Books*.

Amicus Journal	"High Noon," Winter 2001.
Bruce D. Berkowitz	"Global Warming Studies Are a Model of Confusion," *Wall Street Journal*, November 4, 1998.
Paul Brown	"Global Warming: Worse than We Thought," *World Press Review*, February 1999.
H. Sterling Burnett	"Scientific Doubts About Global Warming," *Human Events*, January 12, 2001. Available from One Massachusetts Ave. NW, Washington, DC 20001.
Gale E. Christianson	"Naysayers, Thriving in the Heat," *New York Times*, July 8, 1999.
Eugene Linden Churchill	"The Big Meltdown," *Time*, September 4, 2000.
Gregg Easterbrook	"Hot and Not Bothered," *New Republic*, May 4, 1998.
Douglas Gantenbein	"The Heat Is On," *Popular Science*, August 1999.
Kenneth Green	"Heated Debate over a Hot Theory," *World & I*, January 2001. Available from 3600 New York Ave. NE, Washington, DC 20002.
Mark Hertsgaard	"Severe Weather Warning," *New York Times Magazine*, August 2, 1998.
Richard A. Kerr	"Research Council Says U.S. Climate Models Can't Keep Up," *Science*, February 5, 1999.
Bill McKibben	"Too Hot to Handle," *New York Times*, January 5, 2001.
Dolly Setton	"Some Like It Cold," *Forbes*, June 14, 1999.

What Causes Global Warming?

Chapter Preface

The theory of man-made global warming rests on scientific evidence that the combustion of fossil fuels in the service of human energy needs has added high concentrations of carbon dioxide and methane and other heat-trapping greenhouse gases to the earth's atmosphere. In the opinion of international scientists, global temperatures have risen significantly since the onset of the Industrial Revolution in the mid-1800s, when human activity became dependent on the use of coal, oil, and gas.

In fact, many researchers are convinced that human activity is responsible for the majority of the warming that has occurred over the past 150 years and practically all of the warming since the 1970s. A study of the climate of the last thousand years by Thomas J. Crowley, a geologist at Texas A&M University, found that "natural variability plays only a subsidiary role in the 20th-century warming and that the . . . explanation for most of the warming is that it is due to the anthropogenic increase in GHG (greenhouse gases)."

On the other hand, scientists skeptical that humans are responsible for global warming do not necessarily see a causal relationship between energy consumption and rising global temperatures. Instead, they argue that the earth's climate has continually undergone periods of warming and cooling due to natural factors such as the varying radiation of light and heat from the sun. Sallie Baliunas and Willie Soon, research scientists at the Harvard University Department of Astronomy, argue that "most of the warming early in [the twentieth] century . . . must have been due to natural causes of climatic change, and these natural causes must be understood in order to make an accurate assessment of the effect upon climate of any human activities."

Advances in climate technology may soon provide a more definitive picture of natural and human impacts on the earth's temperature. For example, satellite measurements of solar radiation, available only since the 1970s, are adding important new data to the debate. The authors in the following chapter present their opinions on the mechanisms behind global warming.

"The question is not whether climate will change in response to human activities, but rather how much, how fast, and where."

Human Activity Causes Global Warming

Robert T. Watson

In the following viewpoint, Robert T. Watson claims that the overwhelming majority of scientific experts believe that human activity is causing global warming through the emission of greenhouse gases into the atmosphere. Watson maintains that the global warming trend recorded throughout the twentieth century is extreme and atypical, particularly when contrasted with earth's relatively stable climate during the past ten thousand years. Human-induced global warming will put more stress on ecological and economic systems already adversely affected by pollution and poor management, in the author's opinion. Watson is chairman of the Intergovernmental Panel on Climate Change (IPCC), established in 1988 by the United Nations Environment Programme to evaluate the science of climate change.

As you read, consider the following questions:
1. What should government decision-makers realize about carbon dioxide, according to the author?
2. What evidence does Watson give to support his claim that the earth's climate is changing?
3. According to Watson, how rapidly do oceans respond to changes in greenhouse gas concentrations?

Reprinted from Robert T. Watson's presentation at the Sixth Conference of Parties to the United Nations Framework Convention on Climate Change, November 13, 2000, as found at www.ipcc.ch/press/sp-cop6.htm.

O ne of the major challenges facing humankind is to provide an equitable standard of living for this and future
generations: adequate food, water and energy, safe shelter
and a healthy environment (e.g., clean air and water). Unfortunately, human-induced climate change, as well as other
global environmental issues such as land degradation, loss of
biological diversity and stratospheric ozone depletion,
threatens our ability to meet these basic human needs.

The overwhelming majority of scientific experts, whilst
recognizing that scientific uncertainties exist, nonetheless
believe that human-induced climate change is inevitable. Indeed, during the last few years, many parts of the world have
suffered major heat waves, floods, droughts, fires and extreme weather events leading to significant economic losses
and loss of life. While individual events cannot be directly
linked to human-induced climate change, the frequency and
magnitude of these types of events are predicted to increase
in a warmer world.

Reduction and Adaptation

The question is not whether climate will change in response
to human activities, but rather how much (magnitude), how
fast (the rate of change) and where (regional patterns). It is
also clear that climate change will, in many parts of the
world, adversely affect socio-economic sectors, including water resources, agriculture, forestry, fisheries and human settlements, ecological systems (particularly forests and coral
reefs), and human health (particularly diseases spread by insects), with developing countries being the most vulnerable.
The good news is, however, that the majority of experts believe that significant reductions in net greenhouse gas emissions are technically feasible due to an extensive array of
technologies and policy measures in the energy supply, energy demand and agricultural and forestry sectors. In addition, the projected adverse effects of climate change on socio-
economic and ecological systems can, to some degree, be
reduced through proactive adaptation measures. These are
the fundamental conclusions, taken from already approved/
accepted Intergovernmental Panel on Climate Change
(IPCC) assessments, of a careful and objective analysis of all

relevant scientific, technical and economic information by thousands of experts from the appropriate fields of science from academia, governments, industry and environmental organizations from around the world.

Decision-makers should realize that once carbon dioxide, the major anthropogenic greenhouse gas, is emitted into the atmosphere, it stays in the atmosphere for more than a century. This means that if policy formulation waits until all scientific uncertainties are resolved, and carbon dioxide and other greenhouse gases are responsible for changing the Earth's climate as projected by all climate models, the time to reverse the human-induced changes in climate and the resulting environmental damages, would not be years or decades, but centuries to millennia, even if all emissions of greenhouse gases were terminated, which is clearly not practical.

This viewpoint, which briefly describes the current state of understanding of the Earth's climate system and the influence of human activities . . . is based on accepted and approved conclusions from the IPCC Second Assessment Report (SAR) and a series of Technical Papers and Special Reports . . . and supplemented by recent information that is being assessed in the draft IPCC Third Assessment Report (TAR).

The Earth's Climate System: The Influence of Human Activities

The Earth's climate is changing: The Earth's climate has been relatively stable since the last ice age (global temperature changes of less than 1 degree Centigrade over a century during the past 10,000 years). During this time modern society has evolved, and, in many cases, successfully adapted to the prevailing local climate and its natural variability. However, the Earth's climate is now changing. The Earth's surface temperature . . . is clearly warmer than any other century during the last thousand years, i.e., the climate of the 20th century is clearly atypical. The Earth has warmed by between 0.4 and 0.8 degree Centigrade over the last century, with land areas warming more than the oceans, and with the last two decades being the hottest of the twentieth century (see graph). Indeed, the three warmest years during the last one hundred years have all occurred in the 1990s and the twelve

warmest years during the last one hundred years have all occurred since 1983. In addition, there is evidence that precipitation patterns are changing, that sea level is increasing, that
glaciers are retreating world-wide, that Arctic sea ice is thinning, and that the incidence of extreme weather events is increasing in some parts of the world.

⨴ *The atmospheric concentrations of greenhouse gases are changing due to human activities:* The atmospheric concentrations
of greenhouse gases have increased because of human activities, primarily due to the combustion of fossil fuels (coal, oil
and gas), deforestation and agricultural practices, since the
beginning of the pre-industrial era around 1750: carbon
dioxide by nearly 30%, methane by more than a factor of
two, and nitrous oxide by about 15%. Their concentrations
are higher now than at any time during the last 420,000
years, the period for which there are reliable ice-core data,
and probably significantly longer. In addition, the combustion of fossil fuels has also caused the atmospheric concentrations of sulfate aerosols to have increased. Greenhouse
gases tend to warm the atmosphere and, in some regions,
primarily in the Northern Hemisphere, aerosols tend to cool
the atmosphere.

*The weight of scientific evidence suggests that the observed
changes in the Earth's climate are, at least in part, due to human
activities:* Climate models that take into account the observed
increases in the atmospheric concentrations of greenhouse
gases, sulfate aerosols and the observed decrease in ozone in
the lower stratosphere, in conjunction with natural changes
in volcanic activity and in solar activity, simulate the observed changes in annual mean global surface temperature
quite well. This, and our basic scientific understanding of
the greenhouse effect, suggests that human activities are im-
plicated in the observed changes in the Earth's climate. In
fact, the observed changes in climate, especially the increased temperatures since around 1970, cannot be explained by changes in solar activity and volcanic emissions
alone. The observed changes in temperature, especially
those since around 1970, can be simulated quite well by a climate model that takes into account human-induced changes
in greenhouse gases and aerosols. Not only is there evidence

of a change in climate at the global level consistent with climate models, but there is observational evidence of regional changes in climate that are consistent with those predicted by climate models. For example, climate models predict an increase in intense rainfall events over the United States of America consistent with the observations.

Rising Concentrations of Carbon Dioxide

Emissions of greenhouse gases are projected to increase in the future due to human activities: Future emissions of greenhouse gases and the sulfate aerosol precursor, sulfur dioxide, are sensitive to the evolution of governance structures worldwide, changes in population and economic growth, the rate of diffusion of new technologies into the market place, energy production and consumption patterns, land-use practices, energy intensity, and the price and availability of energy. While different development paths can result in quite different greenhouse gas emissions, most projections suggest that greenhouse gas concentrations will increase significantly during the next century in the absence of policies specifically designed to address the issue of climate change (IPCC Special Report on Emission Scenarios—SRES). Some projections suggest that an initial increase in emissions could be followed by a decrease after several decades if there was a major transition in the world's energy system due to the pursuit of a range of sustainable development goals. The SRES reported, for example, carbon dioxide emissions from the combustion of fossil fuels are projected to range from about 5 to 35 GtC (gross tons carbon) per year in the year 2100: compared to current emissions of about 6.3 GtC per year. Such a range of emissions would mean that the atmospheric concentration of carbon dioxide would increase from today's level of about 365 ppmv (parts per million by volume) to between about 550 and 1000 ppmv by 2100.

Latest projections of carbon dioxide emissions are consistent with earlier projections, but projected sulfur dioxide emissions are much lower: While the SRES reported similar projected energy emissions for carbon dioxide to the 1992 projections, it differed in one important aspect from the 1992 projections, inso-far-as the projected emissions of sulfur dioxide are much lower, because of structural changes in the energy system and because of concerns about local and regional air pollution (i.e., acid deposition). This has important implications for future projections of temperature changes, because sulfur dioxide emissions lead to the formation of sulfate aerosols in the atmosphere, which as stated earlier can partially offset the warming effect of the greenhouse gases.

Global Observed Temperatures

Combined global land, air, and sea surface temperatures 1860 to August 1998 (relative to 1961–1990 average)

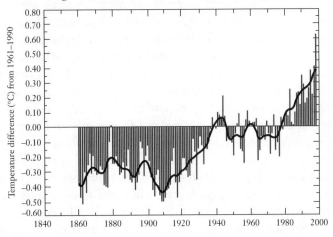

The U.K. Meteorological Office, *Climate Change and Its Impacts: A Global Perspective*, 1997.

Global mean surface temperatures are projected to increase by about 1.5 to 6.0°C (degrees Centigrade) *by 2100.* Based on the range of climate sensitivities and the plausible ranges of greenhouse gas and sulfur dioxide emissions reported in the SRES, a number of climate models project that the global mean surface temperature could increase by about 1.5 to 6.0°C by 2100. This range compares to that reported in the IPCC SAR of 1.0–3.5°C. The revised higher estimates of projected warming arise because the lower projected emissions of sulfur dioxide result in less offset of the warming effect of the greenhouse gases. These projected global-average temperature changes would be greater than recent natural fluctuations and would also occur at a rate significantly faster than observed changes over the last 10,000 years. Temperature changes are expected to differ by region with high latitudes projected to warm more than the global average, and during the next century land areas are projected to warm more than the oceans, and the Northern Hemisphere is projected to warm more than the Southern Hemisphere. However, the reliability of regional scale predictions is still low.

Projections for a Changing Climate

Seasonal and latitudinal shifts in precipitation with arid and semi-arid areas becoming drier: Model calculations show that evaporation will be enhanced as the climate warms, and that there will be an increase in global mean precipitation and an increase in the frequency of intense rainfall. However, not all land regions will experience an increase in precipitation, and even those land regions with increased precipitation may experience decreases in run-off and soil moisture, because of enhanced evaporation. Seasonal shifts in precipitation are also projected. In general, precipitation is projected to increase at high latitudes in winter, while run-off and soil moisture is projected to decrease in some mid-latitude continental regions during summer. The arid and semi-arid areas in Southern and Northern Africa, Southern Europe, the Middle East, parts of Latin America and Australia are expected to become drier.

Sea level projected to rise about 15–95 centimeters (cms) by 2100: Associated with changes in temperature, sea level is projected to increase by about 15–95 cm by 2100 (IPCC SAR), caused primarily by thermal expansion of the oceans and the melting of glaciers. The revised temperature projections are not likely to result in significantly different projections of changes in sea level over the next century because of the large thermal inertia of the oceans, i.e., the temperature of the oceans responds very slowly to a change in greenhouse gas concentrations. However, recent more advanced models are tending to project somewhat lower values of sea level rise. It should be noted that even when the atmospheric concentrations of greenhouse gases are stabilized, temperatures will continue to increase by another 30–50% over several decades, sea level will continue to rise over hundreds of years and ice sheets will continue to adjust for thousands of years. . . .

The frequency and magnitude of ENSO events may increase: Long-term, large-scale, human-induced changes in climate are likely to interact with natural climate variability on time-scales of days to decades (e.g., the El Nino–Southern Oscillation (ENSO) phenomena). Recent trends in the increased frequency and magnitude of ENSO events, which lead to severe floods and droughts in regions of the tropics and subtropics, are projected to continue in many climate models.

Incidence of some extreme events projected to increase: While the incidence of extreme temperature events, floods, droughts, soil moisture deficits, fires and pest outbreaks is expected to increase in some regions, it is unclear whether there will be changes in the frequency and intensity of extreme weather events such as tropical storms, cyclones, and tornadoes. However, even if there is no increase in the frequency and intensity of extreme weather events there may be shifts in their geographic location to places less prepared and more vulnerable to such events. . . .

There are a number of general conclusions that can be easily drawn: (i) human-induced climate change is an important new stress, particularly on ecological and socio-economic systems that are already affected by pollution, increasing resource demands, and non-sustainable management practices; (ii) the most vulnerable systems are those with the greatest sensitivity to climate change and the least adaptability; (iii) most systems are sensitive to both the magnitude and rate of climate change; (iv) many of the impacts are difficult to quantify because existing studies are limited in scope; and (v) successful adaptation depends upon technological advances, institutional arrangements, availability of financing and information exchange, and that vulnerability increases as adaptation capacity decreases. Therefore, developing countries are more vulnerable to climate change than developed countries. . . .

A Time for Action

Significant reductions in greenhouse gases can be accomplished by pursuing sustainable development goals. A future world with greenhouse gas emissions comparable to those of today can either be achieved through the adoption of specific polices, practices and technologies to limit greenhouse gas emissions or through the adoption of a range of policies, practices and technologies to achieve other sustainable development goals (SRES). It should be noted that a major oil company, Shell, has suggested that the mix of energy sources could change radically during the next century. Non-fossil energy sources (solar, wind, modern biomass, hydropower, geothermal and nuclear) could account for as much as half of all energy produced by the middle of this century. Such a future would be consistent with the lower projections of greenhouse gas emissions and would clearly eliminate the highest projections of greenhouse gases from being realized. However, an energy efficient and low-carbon energy world is considered by many to be unlikely to occur without significant policy reform, technology transfer, capacity-building and enhanced public and private sector energy research and development programs. . . .

If actions are not taken to reduce the projected increase in greenhouse gas emissions, the Earth's climate is projected to change at a rate unprecedented in the last 10,000 years with adverse consequences for society, undermining the very foundation of sustainable development.

Policymakers are faced with responding to the risks posed by anthropogenic emissions of greenhouse gases in the face of significant scientific uncertainties. They may want to consider these uncertainties in the context that climate-induced environmental changes cannot be reversed quickly, if at all, due to the long time scales (decades to millennia) associated with the climate system. Decisions taken during the next few years may limit the range of possible policy options in the future because high near-term emissions would require deeper reductions in the future to meet any given target concentration. Delaying action would increase both the rate and the eventual magnitude of climate change, and hence adaptation and damage costs. . . .

Uncertainty does not mean that a nation or the world community cannot position itself better to cope with the broad range of possible climate changes or protect against potentially costly future outcomes. Delaying such measures may leave a nation or the world poorly prepared to deal with adverse changes and may increase the possibility of irreversible or very costly consequences. Options for mitigating change or adapting to change that can be justified for other reasons today and make society more flexible or resilient to anticipated adverse effects of climate change appear particularly desirable.

"Far from being a self-induced disaster, global warming is the result of natural changes in the Earth's climate that promises to yield humanity positive benefits."

Natural Factors Cause Global Warming

John Carlisle

In the following viewpoint, John Carlisle contends that the global temperature increase of 1.5 degrees Fahrenheit over the last 150 years has been caused by natural fluctuations in the earth's temperature and has not been the result of human activity. Carlisle describes periods from the earth's geologic history that had temperatures warmer than the present average of 59 degrees Fahrenheit, illustrating a continually changing climate system. The author believes that humanity will benefit from a warmer planet, enjoying longer growing seasons and higher-yield harvests. Carlisle is the director of the Environmental Policy Task Force of the National Center for Public Policy Research, a conservative/free market foundation in Washington, D.C.

As you read, consider the following questions:

1. How long do the interglacial periods following ice ages typically last, according to Carlisle?
2. What was the average temperature variation in many of the warming and cooling periods of the Holocene, in Carlisle's opinion?
3. According to the author, what is the significance of the Little Ice Age in relation to man-made global warming?

Reprinted, with permission, from "Global Warming: Enjoy It While You Can," by John Carlisle, *National Center for Public Policy Research Policy Analysis*, no. 194, April 1998.

Policymakers have been arguing for nearly a decade over what to do about global warming. Noticeably missing from this debate has been any mention of the fact that natural fluctuations in the Earth's temperature, not Man, is the likely explanation for any recent warming.

Proponents of the global warming theory repeatedly cite a 1.5° F (Fahrenheit) temperature increase over the last 150 years as evidence that man-made CO_2 (carbon dioxide) is dangerously heating up the planet and will cause huge flooding, severe storms, disease and a mass exodus of environmental refugees. Based on this, the Clinton Administration and its environmental allies wanted Congress to ratify a treaty that will hike consumer prices 40 percent and cost the American economy $3.3 trillion over 20 years. But the apocalyptic predictions on which they justified these drastic steps are totally unsubstantiated and ignore some fundamental truths about the Earth's climatic behavior.

The fact is, the planet's temperature is constantly rising and falling. To put the current warming trend in perspective, it's important to understand the Earth's geological behavior.

Glaciation Cycles and Climate Change

Over the last 700,000 years, the climate has operated on a relatively predictable schedule of 100,000-year glaciation cycles. Each glaciation cycle is typically characterized by 90,000 years of cooling, an ice age, followed by an abrupt warming period, called an interglacial, which lasts 10,000–12,000 years. The last ice age reached its coolest point 18,000 to 20,000 years ago when the average temperature was 9–12.6° F cooler than present. Earth is currently in a warm interglacial called the Holocene that began 10,700 years ago.

Although precise temperature readings over the entire period of geologic history are not available, enough is known to establish climatic trends. During the Holocene, there have been about seven major warming and cooling trends, some lasting as long as 3000 years, others as short as 650. Most interesting of all, however, is that the temperature variation in many of these periods averaged as much as 1.8° F, .3° F more than the temperature increase of the last 150 years. Furthermore, of the six major temperature variations occurring prior to the current era, three produced temperatures warmer than the present average temperature of 59° F while three produced cooler temperatures.

For example, when the Holocene began as the Earth was coming out of the last Ice Age around 8700 B.C., the average global temperature was about 6° F cooler than it is today. By 7500 B.C., the climate had warmed to 60° F, 1° F warmer than the current average temperature. However, the temperature fell again by nearly 2° F over the next 1,000 years, settling at an average of 1° F cooler than the current climate.

Warmer Periods Brought Prosperity

Between 6500 and 3500 B.C., the temperature increased from 58° F to 62° F. This is the warmest the Earth has been during the Holocene, which is why scientists refer to the period as the Holocene Maximum. Since the temperature of the Holocene Maximum is close to what global warming models project for the Earth by 2100, how Mankind faired during the era is instructive. The most striking fact is that it was during this period that the Agricultural Revolution began in the Middle East, laying the foundation for civilization. Yet, Greenhouse theory proponents claim the planet will experience severe environmental distress if the climate is that warm again.

Since the Holocene Maximum, the planet has continued to experience temperature fluctuations. In 900 A.D. the planet's temperature roughly approximated today's temperature. Then, between 900 and 1100 the climate dramatically warmed. Known as the Medieval Warm Period, the temperature rose by more than 1° F to an average of 60° or 61° F, as much as 2° F warmer than today. Again, the temperature during this period is similar to Greenhouse predictions for 2100, a prospect global warming theory proponents insist should be viewed with alarm. But judging by how Europe prospered during this era, there is little to be alarmed about. The warming that occurred between 1000 and 1350 caused the ice in the North Atlantic to retreat and permitted Norsemen to colonize Iceland and Greenland. Back then, Greenland was actually green. Europe emerged from the Dark

Reconstruction of the Earth's Climate over the Most Recent 850,000 Years

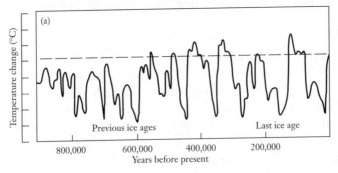

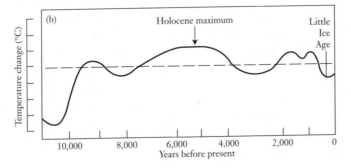

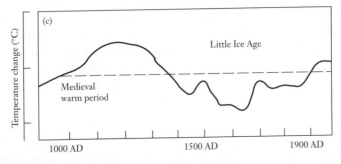

A long-range view of climate indicates that the Holocene, our current period of interglacial climate, is due to end and a period of 90,000 years of cooling is due to begin, taking us back to temperatures 5° to 7°C cooler than those of the present. This schematic shows the mean global temperature for the past million years (a), and in more detail for the past 10,000 years (b), and the past 1,000 years (c). There were at least three cyclic warmings and coolings in the past 10,000 years, lasting about 2,500 years each. The dashed line represents conditions near the beginning of the 20th century.

Source: IPCC, 1990, Figure 7.1.

Ages in a period that was characterized by bountiful harvests and great economic prosperity. So mild was the climate that wine grapes were grown in England and Nova Scotia.

Recovering from an Ice Age

The major climate change that followed the Medieval Warm Period is especially critical as it bears directly on how to assess our current warming period. Between 1200 and 1450, the temperature plunged to 58° F. After briefly warming, the climate continued to dramatically get colder after 1500. By 1650, the temperature hit a low of 57° F. This is regarded as the coldest point in the 10,000-year Holocene geological epoch. That is why the era between 1650 and 1850 is known as the Little Ice Age. It was during this time that mountain glaciers advanced in Switzerland and Scandinavia, forcing the abandonment of farms and villages. Rivers in London, St. Petersburg and Moscow froze over so thoroughly that people held winter fairs on the ice. There were serious crop failures, famines and disease due to the cooler climate. In America, New England had no summer in 1816. It wasn't until 1860 that the temperature sufficiently warmed to cause the glaciers to retreat.

The significance of the Little Ice Age cannot be overestimated. The 1.5° F temperature increase over the last 150 years, so often cited as evidence of man-made warming, most likely represents a return to normal temperatures following a 400-year period of unusually cold weather. Even the United Nation's Intergovernmental Panel on Climate Change (IPCC), the chief proponent of the Kyoto Protocol global warming treaty signed in December 1997, concludes that: "The Little Ice Age came to an end only in the nineteenth century. Thus, some of the global warming since 1850 could be a recovery from the Little Ice Age rather than a direct result of human activities."

Leading climate scientist Dr. Hugh Ellsaesser of the Lawrence Livermore National Laboratory says we may be in for an additional 1.8° F of warming over the next few centuries, regardless of Man's activities. The result would be warmer nighttime and winter temperatures, fewer frosts and longer growing seasons. Since CO_2 stimulates plant growth and lessens the need for water, we could also expect more bountiful harvests over the next couple of centuries. This is certainly not bad news to the developing nations of the world struggling to feed their populations.

Thus, far from being a self-induced disaster, global warming is the result of natural changes in the Earth's climate that promises to yield humanity positive benefits. In the geological scheme of things, the warming is not even that dramatic compared to the more pronounced warming trends that occurred during the Agricultural Revolution and the early Middle Ages. Moreover, there is strong evidence that this long-needed warming is moderating. All things considered, global warming should be viewed for what it is: A gift from the often fickle force of Nature. Enjoy it while you can.

"If indeed there has been warming [in the last 100 years], it is clear the Sun did it, not mankind or greenhouse gases."

Solar Variability Causes Global Warming

John L. Daly

John L. Daly, the author of *The Greenhouse Trap* and numerous articles on climate change, argues in the following viewpoint that the 0.5 degree Centigrade (1.5 degrees Fahrenheit) temperature increase recorded over the past century is due almost entirely to varying radiation from the sun. According to Daly, the last 120 years was a period of intense solar activity, with the biggest radiation peak ever recorded occurring in 1980, resulting in a hotter sun and rising global temperatures. Daly claims that global warming proponents downplay the role of the sun because admitting to solar variability would release humans from blame for rising temperatures.

As you read, consider the following questions:

1. How often does the sun's radiation peak, according to the author?
2. According to Daly, what is misleading about recent claims that the earth is now warmer than at any time in the last six hundred years?
3. In the author's opinion, what caused the "Dickensian Winters" of the nineteenth century?

Reprinted, with permission, from "Days of Sunshine," by John L. Daly, found at www.microtech.com/au/daly/solar.htm.

While the greenhouse industry, led by the Intergovern-mental Panel on Climate Change (IPCC), claims there has been a global warming of around +0.5 degree Centigrade (deg.C) during the twentieth century, this is based entirely on data from a highly flawed surface network of thousands of measuring stations, mostly located in cities. While there may have been some warming, it is unlikely to be as much as the half degree claimed. . . . However, for the purposes of this discussion, it will be assumed there has been some global warming over the last 120 years, by an amount up to +0.5 deg.C, even though +0.25 deg.C would be a more realistic figure.

Given that assumption, the question then arises as to what caused it. Was it just natural variation? Was mankind to blame due to greenhouse gases? Or, was it the primary en-gine of our warmth—the Sun?

Our Variable Star

The Sun appears to us as a steady brilliant light in the sky. Its temperature is nearly 6,000 deg.C and its radiation of light and heat across the vast distances of space weakens in proportion to the inverse square of that distance. By the time solar energy reaches the Earth, the radiation amounts to 1,368 watts per square metre (w/m^2).

This figure is known by solar physicists as the "Solar Constant"

Q.—When is a constant not a constant?

A.—When it's the Solar Constant!

The solar "constant" is actually a variable. Contradictory as this may sound, it is only fairly recently that the radiation from the Sun has been proved to vary according to a highly predictable cycle. The variations so far observed from satel-lites are not much—a mere 0.2% change in mean radiation (or about 2 w/m^2).

Extreme short-term variations have been recorded of up to 0.4%. The Sun, in effect, behaves more like a flickering can-dle, and has been accurately measured doing this by various satellites, including the ill-fated "Solar Max" satellite whose 9-year sample of solar radiation from 1980 to 1989 proved conclusively that the Sun was indeed a variable star. . . .

Further evidence of the Sun's variability came from the Shanghai observatory of the Chinese Academy of Sciences in April 1990, when they reported that the Sun's radius had shrunk 410 kilometres from 1715 to 1987, based on solar eclipse studies. A shrinking sun since 1715 would result in a hotter solar temperature, and increased solar radiation, consistent with a warming of the Earth as claimed by the IPCC.

The 11-Year Solar Cycle

We can measure the variability of the Sun by either direct measurement of its radiation, as done by "Solar Max" and "Nimbus-7", or by its emission of microwave energy. Such observations reveal that the Sun goes through a regular cycle of activity, with its radiation peaking every 11 years, an event known as the Solar Maximum. During periods in between these peaks of activity, the sun cools slightly, a phase in the cycle known as the Solar Minimum. These cycles are quite consistent, and are predictable to within a year. The last maximum peaked in June 1989 and extend well into 1991. The next maximum was predicted by Australia's Ionospheric Prediction Service to peak in May 2000.

During the Solar Maximum, the sun breaks out in sunspots, a direct indication of more intense solar radiation and activity. Although the sunspots are darker than the rest of the Sun, this does not result in less radiation as one might think, but results in the rest of the Sun being hotter. Counting the number and size of sunspots is therefore an indirect measure of how active the Sun is, rather like the severity of blemishes on the skin being an indirect measure of general health.

Although we did not have the benefit of satellites before the 1970's to measure solar radiation, we do have accurate data on sunspot counts going back to around 1600 AD. . . . The intensity of a Solar Maximum is not the same from one cycle to the next, nor are the cycles themselves of exactly the same duration. Eleven years is only an average, but some cycles are as little as 10 years, while some others may extend to 12 years.

Sunspot Counts and Global Temperature

The most remarkable event in the last 500 years was the "Maunder Minimum", a 50-year period from 1650 to 1700 when there were hardly any sunspots at all, indicating a pro-longed cooler period on the Sun. Another period from 1800 to 1830 shows very reduced activity, and a further moderate period occurred in the late 19th and early 20th century. Contrast these with the heightened level of activity and warmth in the latter half of the 20th century, the most intense since solar observations began around 1600 AD.

It is possible to estimate global temperature back to 1600 AD based on indirect proxy measures such as tree ring widths, pollens, oxygen isotopes in ice cores, etc., and when we combine estimates of global temperature and the average of sunspot counts since 1600 AD, we find a close relationship between the increasing solar activity from its low point in the Little Ice Age around 1600 AD and estimated global temperature.

Recent claims that the Earth is now warmer than at any time in the last 600 years are actually quite true, but this is merely a misleading statistical sleight of hand hiding the fact that global temperature was at a historic low point 600 years ago. If we go back in time a little further, say, to 800 years ago, we then find we are starting at the high point of the millennium (the "Medieval Optimum"), resulting in our period being cooler than it was during the time of the Vikings. It's all a question of the year you choose to start from. If you start 600 years ago during the Sporer Minimum, then there has been a long term warming since then. But starting 800 years ago, we find a cooling. Selective use of the start date of any data series is the oldest statistical trick in the book.

Solar Activity and "Dickensian Winters"

The Maunder Minimum of solar activity during the 17th century occurred at exactly the same time as the Little Ice Age. The low level of solar activity in the period 1800–1830 coincided with another cool climatic period dubbed the "Dickensian Winters" (Charles Dickens was a young boy at the time, and his novels depicting snowy Christmases in London, which normally does not get snow in December, reflect his memories of childhood). 1816 has been called "The Year Without a Summer", due to the severe cold which affected America and Europe that year. In 1814, a frost fair was held on the River Thames in London, indicating that temperatures had very briefly descended to even the Little Ice Age level when Thames frost fairs were common.

The late 19th century was another cooler period, exacerbated by the Krakatoa eruption of 1883. Since that time, the Sun has grown steadily more active, with the biggest Solar Maxima ever recorded peaking in 1957 (based only on sunspot counts), or 1980 (when based on modern satellite and microwave measurement). According to Richard C. Willson and Hugh S. Hudson, the Sun exhibited a "remarkable irradiance excess during 1980, at about the time of the sunspot maximum of solar cycle 21". It is hardly any coincidence that global climate warmed up at exactly the same time, as measured by satellites (the most accurate measure of global temperature).

The Solar Cycle

Variations in solar irradiance 1900–1990 (in watts per square metre)

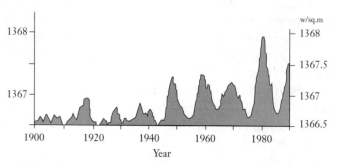

1980 saw the biggest maximum ever recorded, followed by 1990 as 2nd biggest.

John L. Daly, "Days of Sunshine."

Since the cooler Sun was clearly responsible for the Little Ice Age, and the Dickensian Winters, then the hotter Sun of the latter half of the 20th century would naturally create a warmer climate. This is consistent with the view that some warming has occurred.

The Little Ice Age (and Maunder Minimum which caused it) do present one puzzle though, namely, that the Little Ice Age was more severe than would be justified by a –0.2% reduction in solar radiation alone. We know there was a cooling of about 1 degree from today's average, and yet we also know that a 0.2% reduction of solar radiation during the Maunder Minimum would not have been enough in itself to cause such a severe climate downturn.

The 0.2% solar cooling estimate is only based on recent satellite observations, covering only 17 years. Over a longer time span, variability in the Sun may very likely exceed this level. Indeed, observations by Lockwood & Skiff of 36 nearby stars similar in size and makeup to our own Sun, revealed that nearly half of them varied by more than 0.5% in brightness over a time interval of less than 4 years. If our sun behaves in a similar way over longer periods, then the severity of the Little Ice Age and the Dickensian Winters period becomes explainable exclusively in terms of solar variations.

Little Room for the Greenhouse Effect

So, how much is the Sun responsible for the 20th century warming?

If we assume a solar radiation increase of only +0.2% in recent years only, (or a 2 w/m² increase in solar radiation), . . . 20th century warming should be about +0.15 deg.C. However, the evidence of the Little Ice Age suggests secondary effects which combine to cause a temperature change up to 3 times as great. Using this criteria, the 20th century warming would therefore be—+0.15 x 3 = +0.45 deg.C.

This is only slightly less than the +0.5 deg warming claimed by the IPCC. Since, unlike Greenhouse, the Sun's impact on climate is certain, there is little or no room left here for the Greenhouse Effect. It is for this reason that Greenhouse promoters downplay the role of the Sun, or more absurdly, even to deny that the sun has any role at all. They even persist in the fiction of referring to the variable level of solar radiation as "the solar constant". For them to admit to solar variability would deny Greenhouse any role at all in what little warming there has been in the 20th century.

The graph shows variations in direct solar radiation during the 20th century, with a very obvious extreme peak in 1980. This is based on a composite of measures developed by Foukal & Lean. The intense solar peak of 1980 is associated with a sharp warming of climate between 1976 and

1980, as is the 1957 event. The solar minimum years, such as 1976, are associated with cooler climate.

Since the oceans are able to absorb and buffer the immediate short-run impact of most solar maxima, short-term correlations between temperature and solar maxima show up most readily when the maximum is a very large one, as in 1957, 1980, and 1990. However, weaker maxima may be masked by other climatic processes, such as the El Nino Southern Oscillation. In the longer term, the heat retention capacity of the oceans would result in long-term climatic warming in reaction to a succession of several major solar maxima, as shown by the last four intense solar maxima. . . .

Some solar physicists predict that the Sun will soon enter another cooler phase, similar to the Maunder Minimum or Dickensian periods. It is inconceivable that the Sun can maintain the high levels of activity experienced during the last four solar cycles indefinitely into the future. This would suggest future global cooling, making discussion about Greenhouse somewhat academic.

This explanation for the claimed warming in the 20th century is more credible, given the Sun is our heat source, whereas natural Greenhouse gases can only recycle existing heat.

The Earth has certainly warmed in the last 100 years, although not as much as suggested by the IPCC. However, the Sun has also been hotter during the same period, so that some, if not all, the warming is attributable to the more intense solar activity of the 20th century and to the shortening of the solar cycles themselves.

If indeed there has been warming, it is clear the Sun did it, not mankind or Greenhouse gases.

*"The rapid warming since 1970 is several
times larger than that expected from any
known or suspected effects of the Sun, and
may already indicate the growing influence
of atmospheric greenhouse gases on the
Earth's climate."*

Solar Variability Has Not
Caused Recent Global Warming

Judith Lean and David Rind

In the following viewpoint, Judith Lean and David Rind assert
that the role of a variable sun in global warming accounts for,
at best, one-half of the recorded warming since 1850 and less
than one-third of the warming observed over the last twenty-
five years. The most likely cause of climate change since the
start of the Industrial Revolution 150 years ago, according to
the authors, is the growing atmospheric concentration of
greenhouse gases originating from human activity. Judith
Lean is a solar physicist at the U.S. Naval Research Laboratory in Washington, D.C. David Rind is a physicist at the
NASA Goddard Institute for Space Studies in New York City.

As you read, consider the following questions:
1. According to the authors, who discovered that sunspots
   come and go in eleven-year cycles?
2. What are the two phenomena that affect the flow of
   energy from the sun, according to Lean and Rind?
3. In the authors' opinion, what effect is the sun expected to
   have on projected temperature increases, based on
   current measurements?

Reprinted from "The Sun and Climate," by Judith Lean and David Rind,
Consequences, vol. 2, no. 1, Winter 1996.

Of the many objects in the universe, only two are essential for life as we know it: the Earth itself, and the Sun: the star around which it circles, year after year. Burning steadily in stable, middle age, the Sun—now about five billion years old—provides an unfailing source of light and energy. The Sun's heat is so intense that at a distance of 93 million miles it warms the surface of the otherwise cold and lifeless Earth some 250° Centigrade, to -18° C (0° Fahrenheit). Thus warmed, the solid Earth releases a portion of its heat in the form of infrared radiation, which is trapped by atmospheric greenhouse gases, further raising the surface temperature to a more comfortable 15° C (59° F). In this way, the Sun's radiation and the Earth's blanket of greenhouse gases sustain the mean global temperature at a level supportive of life. Sunlight also powers photosynthesis, and provides energy for the atmospheric and oceanic circulations that profoundly affect all living things.

A Variable Star

Like other stars of similar age, size, and composition, the Sun shows many signs of variability. Most pronounced and by far the most familiar is a cycle of about eleven years in the number of dark spots on its glowing surface. But although the Sun is known to be a variable star, its total output of radiation is often assumed to be so stable that we can neglect any possible impacts on climate. Testimony to this assumption is the term that has been employed for more than a century to describe the radiation in all wavelengths received from the Sun: the so-called "solar constant," whose value at the mean Sun-Earth distance is a little over 1⅓ kilowatts per square meter of surface.

In truth, the solar "constant" varies. Historical attempts to detect possible changes from the ground were thwarted by variable absorption in the air overhead. Measurements from spacecraft avoid this problem, and the most precise of these, made continuously since 1979, have revealed changes on all time scales—from minutes to decades—including a pronounced cycle of roughly eleven years. Sunspots and other forms of solar activity are produced by magnetic fields, whose changes also affect the radiation that the Sun emits, including its distribution among shorter and longer wavelengths. The most highly variable parts of the Sun's spectrum of radiation are found at the very shortest wavelengths—the ultraviolet (UV) and X-ray region—and in the very longest and far less energetic band of radio waves.

New insights into the variable nature of the Sun have almost always been followed by efforts to find possible impacts on the Earth—chiefly through comparisons with weather and climate records. Initially the quest was not so much a detached inquiry as a determined effort to demonstrate a long-sought hope: that keys found in the cyclic nature of solar behavior might open the doors of down-to-Earth predictions.

In the latter part of the 19th century, there were many claims of new-found connections between sunspots and climate. It began with the announcement by the amateur astronomer Heinrich Schwabe, in 1843, that sunspots come and go in an apparently regular eleven-year cycle. What followed was a flood of reported correlations, not only with local and regional weather but with crop yields, human health, and economic trends. These purported connections—that frequently broke down under closer statistical scrutiny—lacked the buttress of physical explanation and were in time forgotten or abandoned.

After more than a century of controversy, the debate as to whether solar variability has any significant effect on the climate of the Earth remains to be settled, one way or the other. This long unanswered question has of late emerged anew, and with some urgency—in the context of widespread concerns of impending global greenhouse warming. For in order to gauge the possible impacts of anthropogenic greenhouse gases on the present or future climate, we must first know the natural variations on which our own activities are imposed. . . .

How Much Does the Sun's Radiation Vary?

Observations from space reveal that the total radiation from the Sun is continually changing—with variations of up to 0.2 percent from one month to the next. The timing and nature of these shorter-term fluctuations are consistent with the Sun's 27-day period of rotation, and occur because persisting darker—or brighter—areas on the solar surface alter the amount of sunlight received at the Earth.

This "rotational modulation" of solar total radiation is superimposed on a much-longer cycle of about eleven-years (called the Schwabe cycle, after the discoverer of the sunspot cycle) which had an amplitude of about 0.1 percent in the two most recent cycles. The enhanced solar activity that characterized the Sun at the peak of the last two cycles, in 1979–1981 and again in 1989–1991, increased both the overall brightness and the amplitude of the rotational modulation. The nature of these dark and bright structures, and the reason why the Sun is slightly brighter when there are more sunspots, are explained below.

Still, since we have been able to monitor solar total radi-

Estimates of the Amplitudes of Natural and Human-Induced Climate Forcings in the Past 140 Years

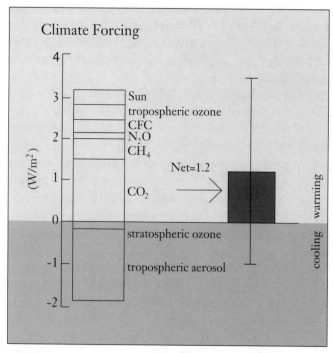

Judith Lean and David Rind, *Consequences*, vol. 2, no. 1, 1996.

ation for only about fifteen years, knowledge of the Schwabe radiation cycle is less than complete. Experimental uncertainties in the spaceborne measurements allow an amplitude as high as 0.15 percent for the Schwabe cycle. Moreover, other cycles may be quite different, as a result of longer-term changes on the Sun. One and a half decades of solar monitoring is simply not long enough to detect other possible cycles of longer period—and perhaps higher amplitude—that may well be fundamental features of the Sun. . . .

Changes in the Sun's total radiation and how it is distributed in wavelength occur primarily because solar activity produces two different phenomena that alter the surface brightness, and hence modulate the outward flow of radiated energy. The first of these are the dark spots that appear in great number during times of high solar activity. Cooler than surrounding regions, sunspots "block" for a time some of the radiation that the Sun would otherwise emit. The second are known collectively as *faculae*. They are brighter than the surrounding surface, and add to the overall radiation from the Sun. The radiation that is emitted from the Sun varies continually in response to the push and pull of these two competing and constantly changing features. In years of maximum solar activity, it is the bright faculae that prevail, raising the levels of both total and UV radiation. . . .

How Sensitive Is the Earth's Climate to Changes in Solar Radiation?

The sensitivity of climate to solar radiation changes, as defined earlier, is not well known. A conservative estimate is that a 0.1 percent change in solar total radiation will bring about a temperature response of 0.06 to 0.2° C, providing the change persists long enough for the climate system to adjust. This could take ten to 100 years. . . .

The most likely cause of climate change in the period since about 1850, based on estimated magnitudes of known perturbations, is the growing concentration of greenhouse gases: in particular carbon dioxide (CO_2), methane (CH_4), nitrous oxide (N_2O), and the commercially-made compounds of chlorine, fluorine, and carbon called *halocarbons*. Other causes, including solar radiative output variations, are thought to be less important, but they are also far less certain. Most poorly known is the magnitude of the potentially-large cooling effect of atmospheric aerosols.

A comparison of significant climate perturbations for the period from 1850 through 1990 is shown (see graph) in terms of the estimated energy that each has added to or subtracted from a stable climate system. Some, like the increasing radiation from the Sun, have added energy; others, like atmospheric aerosols or stratospheric ozone, have taken it away. The net result is an addition of 1.2 watts per square meter. The Sun accounts for about a quarter of the net amount, even though it contributes but 10 percent of the total on the positive, warming side of the balance sheet. Most of the change is attributed to the push and pull of the other, much larger perturbations—which are predominantly of human origin.

If we assume that the climate is equally sensitive to radiative forcing from each of these causes, the net increase of 1.2 watts should have brought about an increase in global mean temperature of 0.3 to 1.1° C, depending on the climate sensitivity that is assumed. The documented rise of about 0.5° C in the same period falls at the low end of this range. It may be premature to make such a comparison, however, since it is uncertain when all of the warming would be felt, given the lag times of up to a century that are imposed on the climate system by the thermal inertia of the oceans.

What Can We Learn from Climate Changes in the Pre-Industrial Era?

What is known of climate change prior to the Industrial Revolution affords additional insight. The well-documented surface temperature rise since 1850 can be viewed as but the most recent 60 percent of a warming of about 0.8° C since the 17th century, interrupted periodically by volcanic effects. Estimates of Northern hemisphere surface temperatures from 1610 to 1800—during part of the so-called Little Ice Age—correlate well with a reconstruction of changes in solar total radiation—around the time of the Maunder Minimum. This suggests, without proving, a predominant solar influence on climate throughout this 200 year, pre-industrial epoch. The reconstructions of solar radiation and surface temperature for these years tell of an increase in solar radiation of 0.14 percent and a coincident warming of 0.28° C. If we apply the same implied sensitivity to the period since 1850, the 0.13 percent increase in solar radiation in the last 140 years should have produced a warming of 0.26° C, or about half of that observed. If we apply the same relationship to the last 25 years, solar changes can account for less than a third of the warming observed.

In summary, extending the relationship between surface warming and solar change from the Maunder Minimum to the present era implies that while the Sun could account for almost all of the temperature changes from 1600 to 1800, it can explain at most, half of the warming from 1600 to the present. When effects of dust from known volcanic eruptions in the early 19th century are also included, solar and volcanic effects can satisfactorily explain much that is known of climate change from 1600 to the beginning of the industrial epoch, around 1850.

How Might Solar Variations Affect Projections of Global Greenhouse Warming?

What has happened in the past can help us assess the relative importance of solar and anthropogenic changes in the climate of the future.

It seems likely that changes in solar radiation, linked to long-term variations in solar activity, may have been the dominant climate driver in the period between about AD 1600 and 1850. As discussed earlier, the explanation of trends in global surface temperature since that time is not as simple, when both the positive and negative impacts of fossil fuel consumption are added to the picture.

Since 1850, variations in global surface temperature appear to track changes in the level of solar activity at least as well as they track increasing greenhouse gas concentrations. At the same time, when probable energy inputs are taken into account (as in graph), solar effects can account for only a fourth of the net change in climate forcing in this 140 year period. It could be that the climate system is more sensitive than we think to changes in solar energy, as opposed to greenhouse gas increases. There could also be errors in our estimates of the magnitudes of other climate forcing terms—especially aerosol cooling—or, as mentioned earlier, time lags in how the climate system responds.

On the basis of what we now know, solar changes might account for a rise of about 0.5° C since the 17th century, perhaps half of the warming since 1850, and less than a third of the warming in the last twenty-five years. The possibility of solar-induced changes of these magnitudes complicates the unambiguous detection of a possible greenhouse warming signal in climate records of the last 100 years or so. We need to remember, however, that to ascribe any significant fraction of the documented warming since 1850 to the Sun requires high estimates of climate sensitivity. It also requires the added assumption that long-term changes in solar total radiation can exceed by two and a half times what has been observed in recent measurements from space.

Looking ahead, were solar changes limited to what has been measured in the last fifteen years, future changes in the Sun's total radiation would have only a negligible effect on the temperature increases of 1 to 3° C that are now projected in IPCC models for the end of the twenty-first century. If greater changes in solar radiation occur—as seems probable based on what is known of climate and solar activity in the past—the Sun needs to be considered in long-term climate projections. The present high levels of solar activity may be approaching all that the Sun can deliver, in terms of total radiation. But, were the Sun's activity and total radiation to drop in the coming century to levels of the Maunder Minimum, solar effects might reduce the expected surface temperature effects of enhanced greenhouse warming—by at most about 0.5° C. . . .

Were that to happen, the principal impacts would be to cloud for a time the unambiguous detection of enhanced greenhouse warming and to soften, by at most about a half, its projected impact on the temperature of the planet. The rapid warming since 1970 is several times larger than that expected from any known or suspected effects of the Sun, and may already indicate the growing influence of atmospheric greenhouse gases on the Earth's climate.

Periodical Bibliography

The following articles have been selected to supplement the diverse views presented in this chapter. Addresses are provided for periodicals not indexed in the *Readers' Guide to Periodical Literature*, the *Alternative Press Index*, the *Social Sciences Index*, or the *Index to Legal Periodicals and Books*.

Robert Adler	"Don't Blame the Sun," *New Scientist*, May 6, 2000. Available from PO Box 4928, Oak Brook, IL 60522.
Wolf Berger	"Global Warming Is Here to Stay, Get Used to It," *San Diego Union-Tribune*, January 19, 2001.
H. Sterling Burnett	"Myths of Global Warming," *Human Events*, June 27, 1997. Available from One Massachusetts Ave. NW, Washington, DC 20001.
Bill Corporon	"Taking the Earth's Temperature," *Lamp*, Fall 1998. Available from 5959 Las Colinas Blvd., Irving, TX 75039.
Thomas J. Crowley	"Causes of Climate Change over the Past 1000 Years," *Science*, July 14, 2000.
The Economist	"Solar Rash and Earthly Fever," February 21, 1998.
Hugh W. Ellsaesser	"What Man-Induced Climate Change?" *21st Century Science & Technology*, Summer 1997. Available from PO Box 16285, Washington, DC 20041.
Richard A. Kerr	"Greenhouse Forecasting Still Cloudy," *Science*, May 16, 1997.
Patrick J. Michaels, Paul C. Knappenberger, and Robert E. Davis	"The Way of Warming," *Regulation*, vol. 23, no. 3, 2000. Available from 1000 Massachusetts Ave. NW, Washington, DC 20001.
Andrew C. Revkin	"Study Faults Humans for Large Share of Global Warming," *New York Times*, July 14, 2000.
Arthur B. Robinson and Noah Robinson	"Some Like It Hot," *American Spectator*, April 2000.
William K. Stevens	"Global Warming: The Contrarian View," *New York Times*, February 29, 2000.
William K. Stevens	"New Evidence Finds This Is Warmest Century in 600 Years," *New York Times*, April 28, 1998.

Curt Suplee   "Sun Studies May Shed Light on Global
                            Warming," *Washington Post*, October 9, 2000.

Simon F.B. Tett et al.   "Causes of Twentieth-Century Temperature
                            Change Near the Earth's Surface," *Nature*, June
                            10, 1999.

Jim Wilson   "Global Warming Wildcard," *Popular
                            Mechanics*, September 9, 1998.

CHAPTER 3

What Will Be the Effects of Global Warming?

Chapter Preface

Advocates of the man-made global warming theory are convinced that rising temperatures will take a toll on human health and economic security, further damaging an environment rendered fragile by pollution, water shortages, and overpopulation. One effect of global warming that is of particular concern to climate forecasters is the growing prevalence of massive hurricanes, floods, and droughts. Scientists and environmentalists believe that these types of extreme weather events will become increasingly common in a world undergoing rapid warming. Many fear that humanity, particularly in poorer nations, remains unprepared to deal with consequences such as famine and displaced populations. Explains Molly O'Meara, a staff researcher at the environmental research organization the Worldwatch Institute, "Climate change would likely increase the number of people in developing nations at risk of hunger. . . . But if climate change causes famine in one part of the world, then other countries will feel the pain too, as environmental refugees course across their borders."

Contrary to these claims, some scientists argue that certain parts of the world have always been susceptible to floods and droughts, and that it is foolish to blame every bad weather event on global warming. According to George H. Taylor, an atmospheric scientist and the state climatologist for Oregon, "There has been a steady migration of people to more vulnerable coastal areas . . . where even small storms can do major damage. Human encroachment has reduced the ability of natural environmental systems to provide protection from extreme weather." Furthermore, other scientists believe that global warming will lead to longer agricultural growing seasons and a healthier population enjoying a more temperate climate. Says Patrick J. Michaels, a professor of environmental studies at the University of Virginia, "The effects of postwar warming have been benign or beneficial. The growing season has lengthened. . . . And heat-related deaths [are in] decline."

Whether the effects of global warming will benefit the environment and humanity will be determined over the course of many years. The authors in the following chapter debate how global warming may affect the planet.

> *"Linked to the destabilizing effects of global warming, Extreme Weather is characterized by higher (and lower) temperatures, fiercer winds, deadlier floods, [and] longer droughts."*

The Effects of Global Warming Will Be Detrimental

Gar Smith

Gar Smith asserts in the following viewpoint that global warming will destabilize humanity and the environment by causing a dramatic increase in extreme weather (EW). EW events such as deadly floods, droughts, and powerful storms are already wreaking havoc across the globe, according to Smith, and he predicts that uncontrolled global warming will force the resettlement of populations and the redesign of industry, disrupting economies worldwide. Smith is editor in chief of the *Earth Island Journal* and winner of the Alternative Press Award for best scientific and environmental reporting for 1997 and 1998.

As you read, consider the following questions:
1. According to the author, how many deaths were caused by extreme weather events in 1998?
2. What are some of the after-effects suffered by communities struck by natural disasters, in Smith's opinion?
3. In the author's opinion, how has the flooding of rivers killed fish and polluted oceans?

Reprinted, with permission, from "W2K: The Extreme Weather Era," by Gar Smith, *Earth Island Journal*, Summer 2000.

When the date rolled over from midnight 1999 to the first minutes of Year 2000, most computers continued to hum and the world's oil-electric-silicon-based societies breathed a sigh of relief.

True, there were a few glitches (the National Clock in Arlington, Virginia, clicked off the first day of the new millennium as January 1, 19,000), but thanks to an unprecedented $30 billion global fixit program, most banks remained open, most lightbulbs continued to glow and most fuel pumps delivered gasoline on demand.

Yet, as the millennium dawned, millions of homes *were* plunged into darkness. In Europe, Africa and South America, water supplies failed, food and fuel deliveries stopped. Nuclear powerplants were dangerously disrupted. Airplanes even fell from the sky.

Global Warming and Extreme Weather Events

This global cataclysm was not caused by computer circuits misinterpreting a binary code. It was caused by a new kind of threat that climatologists are calling Extreme Weather Events. Unlike Y2K (which merely threatened computer-dependent infrastructure), Extreme Weather Events have the power to destroy bridges, bury roads, collapse buildings, obliterate forests and kill people by the thousands.

Linked to the destabilizing effects of global warming, Extreme Weather (EW) is characterized by higher (and lower) temperatures, fiercer winds, deadlier floods, longer droughts, and an increased frequency of dust storms, tsunamis, storm surges, tornadoes, hurricanes and cyclones.

On March 10, 2000, the National Oceanic and Atmospheric Administration (NOAA) announced that the winter of 1999–2000 was the warmest winter since the US began keeping records 105 years ago.

At the International Decade for Natural Disaster Reduction in July 1999, World Meteorological Organization Secretary-General Godwin O.P. Obasi stated that weather-related disasters "are costing the world economy about $50 billion per annum. These disasters have also caused suffering to more than two billion people since 1965 and three million have lost their lives."

In its *Global Environmental Outlook 2000* study, the UN Environmental Program (UNEP) concluded that it is no longer possible to prevent "irreversible harm" to the tropical rainforests. UNEP foresees a number of "full-scale emergencies" on the horizon.

In short, the real threat to civilization has turned out to be not Y2K but W2K—Weather in the Year 2000.

W2K Superstorms Hit Europe

In the last days of the 20th Century, a Pre-Millennium superstorm ripped off roofs and snuffed out lights from Italy to the Netherlands. Tree trunks and mudslides blocked roads. Falling trees, toppled chimneys and collapsing walls crushed and killed more than 120 people.

In Italy, a light plane encountered monstrous winds over Torino province and, yes, fell from the sky.

In the south of France, the storm snapped powerlines to three 900-Mw nuclear powerplants, disabling the emergency cooling systems and forcing an emergency shutdown. The storm also flooded two of the reactors, disabling the pumps used to send water through emergency cooling circuits.

The damage was worst in France where winds as high as 136 mph killed 79 people, uprooted 10,000 trees at Versailles Palace and caused $77 million in damage to French landmarks including Notre Dame Cathedral and Sainte-Chapelle. The French government declared a "natural catastrophe" over two-thirds of the country and mobilized 6,000 troops to clear roads, provide emergency food and water and search for survivors.

"In the meteorological records, there is no trace of a phenomenon as violent as this," marveled French weather service forecaster Hubert Brunet. Two million people lost power. More than 400,000 homes lost telephone service. Millions of homes were damaged. Many towns were without drinking water. Major airports were paralyzed.

W2K's Forest-Killing Winds

The winds that hit Europe and India were forest-killers. In France alone, the blasts flattened 160 square miles of forests and destroyed 400 million trees. The National Forest Office reported a loss equal to three years' worth of timber harvests.

Extreme weather clearcut France's commercial forest industry to the ground. One shaken commercial forester called the storm "the most disastrous event since WWII." The storm also destroyed forests in Germany and Austria.

Since many of the trees were 100–400 years old, it will take centuries to replace what a single extreme weather storm destroyed.

On October 29, 1999, a supercyclone struck Orissa, India, demolishing miles of these newly planted mangrove forests. After an earlier storm devastated the region, the government of Orissa sponsored a major reforestation effort to replace the protective buffer of native mangrove forests. (The mangroves had been cleared to make room for prawn farms.)

The supercyclone was an environmentalist's worst nightmare. One of the working assumptions of restorationists is that massive reforestation can help stabilize the atmosphere while protecting the land from storm damage.

These disasters demonstrate that we now face a new threat—superstorms so powerful that they can destroy hundreds of square miles of trees (old and new) within a matter of hours.

W2K: Summer Blackouts

While floods and hurricanes are dramatic events that can knock out a country's electrical grid, the slow, upward creep of temperatures also can cause blackouts, as over-extended powergrids try to feed air-conditioning demands of exploding urban and suburban populations.

On August 10, 1996, temperatures in California's Central Valley hit 110 F, punching electrical demands to an unprecedented 21,451 megawatts. As the mercury rose, powerlines began to stretch and sag. In Oregon, one of the lines dropped onto a tree limb and the resulting short knocked out power to 4 million people from Canada to Mexico. Five of California's 11 powerplants were forced off-line—including both units of the Diablo Canyon nuclear power station.

Across eight Western states, transformers exploded, gas pumps shut down. ATM machines went blank and power-surges caused fires, destroying homes and garages. People were stuck in elevators, trains, buses, and subway tunnels. Airports, businesses, and supermarkets were disabled. Radar systems crashed, checkout counters went dead, frozen foods thawed and rotted. In Los Angeles, six million gallons of raw sewage was released into the Pacific Ocean, contaminating ten miles of coast.

Power was not fully restored for five days. Even before the Y2K bug surfaced, the lesson was clear: An economy that is dependent on fossil fuels and centralized power sources is fundamentally unstable.

The Economic Impacts

On January 14, UN Environment Program predicted that W2K events "will cause major economic impacts for an insurance industry already burdened with a 14-fold increase in insured losses in the last four decades. The economic losses from the past 24 climate or weather-related catastrophes alone have exceeded $150 billion."

The *Calgary Herald* reports that "financial losses from severe weather events in Canada have increased at 10 times the rate of the country's economic growth since the mid-80s." With insured losses in 1998 hitting $1.45 billion, these rapidly rising disaster costs may soon outrun Canada's ability to finance recovery.

Hurricane Hugo pummeled South Carolina in 1989 causing $47 billion in damage. Three years later, Hurricane Andrew inflicted $30 billion in damage on Florida.

In 1998, Extreme Weather Events—floods, fires and storms—caused 32,000 deaths, left 300 million homeless and cost insurers a record $92 billion. This financial loss does not begin to describe the full impact of these calamities on human society and on the environment.

Untabulated Impacts

"When a community is hit by a major storm," Sheila D. David, Sarah Baish and Betty Hearn Morrow reported in the October 1999 issue of *Environment*, "the entire social fabric that defines a population as a community can be severely weakened. People relocate (some permanently), neighborhoods are destroyed, friendships are severed, support networks are broken, and domestic relationships are stressed. Schools, churches, social groups, and families are apt to never be the same."

Long-term studies show that communities struck by natural disasters suffer increased incidences of suicides, family violence, desertions, and alcohol and drug abuse.

The insured losses do not include harm to ecosystems. Hugo damaged 4.5 million acres (37 percent) of Carolina's forests. Francis Marion National Forest lost 87 percent of the habitat of the endangered red-cockaded woodpecker and 63 percent of the birds.

South America—"The Climate Has Gone Mad"

In May 1999, five northwest Mexican states were declared disaster zones after the longest drought in memory killed crops and cattle. Five months later, torrential rains raked Mexico, unleashing floods that killed at least 400 and left 300,000 homeless. InterPress Service reported that one phrase was heard more than any other in the aftermath of Mexico's drought/flood tribulations: "The climate has gone mad!"

A study from Mexico's National Autonomous University warns that "Mexico will be one of the countries hardest hit by global warming." In 25 years, Mexico may experience 54.5 C temperatures and desperate citizens will fell north to the US and Canada "in search of better climactic and environmental conditions."

In December 1999, floods left as many as 20,000 dead, 35,000 homes demolished and 150,000 homeless in Venezuela. According to the National Oceanic and Atmospheric Administration (NOAA), "Large swatches of Venezuela's northern coast were swept away." Venezuelan President Hugo Chavez toured the area and lamented, "There are bodies in the sea, there are bodies under mud, there are bodies everywhere."

Gar Smith, *Earth Island Journal*, Summer 2000.

After Hugo hit, the white ibis population crashed from 10,000 pairs to zero. More than half of South Carolina's 54 bald eagle breeding sites were obliterated. Saltwater forced inland killed five million catfish, bream, largemouth bass and other fish. All of the unhatched eggs of Loggerhead and four other endangered sea turtles were destroyed along the South Carolina coast (first by the storm and then by subsequent efforts to restore the beach).

It's Already Happening

In 1988, *Earth Island Journal* ran a cover story on the threat of climate change. "There is no advantage in waiting." Environmental Defense Fund Senior Scientist Michael Oppenheimer warned in that issue, "If we don't move fast, there will be so much climate warming that our policy options will be narrowed in the future."

On June 30, 1988, more than 300 scientists from 48 countries called for a 20 percent cut in world oil consumption and a "carbon tax" on fossil fuels. *The Ecologist* called for canceling Third World debt in exchange for "guaranteed protection of the world's remaining tropical forests."

The call went largely unheeded.

Twelve years later, the National Oceanic and Atmospheric Administration (NOAA) is warning that the climate is warming at "an unprecedented rate" and registering changes that were not expected to occur until far into the 21st century.

The polar ice sheets are already melting and rising seas are decimating low-lying islands. As University of Michigan Geologist Henry Pollack told the *Los Angeles Times*, "Even if we don't understand the details of what's causing [climate changes], we still have to deal with the consequences."

A two-year computer modeling study conducted by the Union of Concerned Scientists and the Ecological Society of America foresees West Coast winters become 5 to 6 degrees warmer and summer temperatures rising 1 to 2 degrees over the next 30 to 50 years. The computer models predict that water shortages will devastate the farming economy of the Central Valley while warming ocean waters will deplete populations of plankton, eroding the basis of the marine food chain and triggering die-offs of fish, sea birds and marine mammals.

In regions where rainfall increases, flooding has begun to flush large amounts of nitrogen into rivers, killing fish and eventually carrying deadly effluent into oceans. A report from the US Climate Forum notes that this "disruption of the global nitrogen cycle is happening even faster than the disruption of the global carbon cycle."

NOAA's Hurricane Research Division notes that hurricanes are already increasing in frequency and strength as they pick up energy from warmer ocean surfaces. And each ten percent increase in strength brings a doubling of the damage when the storms crash ashore.

On February 20, 2000, scientists with NOAA's Global Change Research Program (GCRP) informed delegates at the annual meeting of the American Academy for the Advancement of Science in Washington, DC that "human-induced climate change has already started" and was, by now, most likely beyond human control. "You can't stop climate change, given what we're doing right now," said Michael MacCracken, director of the GCRP's National Assessment Coordination Office.

MacCracken's startling announcement was echoed by University of Maryland Scientist Donald Boesch, head of the US National Global Change Assessment. "Things *are* going to happen," Boesch warned. "We're going to have to deal with them."

This view is shared by Randall Hayes, founder of the Rainforest Action Network. Hayes observes that "we have already entered the dangerous era of ECOSPASMS. Hang on for the ride."

An easy transition to the "appropriate-technology, sustainable-society paradigm is an impossibility now," Hayes believes. "We must up-the-ante to ensure that our solution scenarios are on a scale commensurate with the problem.". . .

Facing W2K

Short of an atomic blast, there is no military weapon that wreaks such massive devastation as Extreme Weather. After viewing the onslaught of Hurricane Floyd in 1999, Bill Massey of the Federal Emergency Management Agency (FEMA) Regional Office IV in Atlanta remarked "Now, instead of worrying about the atom bomb, we're worried about bad weather."

In a December 14, 1998 report, the British Parliamentary Office of Science and Technology concluded that "some degree of climate change is inevitable" and urged the UK to prepare for "more frequent and severe extreme weather-related events" that would cause trade disruptions, water shortages, and outbreaks of disease, heatstroke and food poisoning.

The report noted that "climatic zones are likely to shift northwestwards by 50–80 km per decade, so many species of plants and animals would have to migrate to remain in the conditions to which they are suited. Adaptive responses could include providing stepping stones and corridors of appropriate habitat along which mobile species could move." The old conservation strategy of maintaining fixed wildlife "sanctuaries" would doom many species to imprisonment and slow extinction.

Alaska's permafrost is thawing, forcing local and state authorities to spend millions repairing buckled roads and tipping homes. Farmers will be faced with the choice of growing different crops or abandoning their lands.

Surviving the W2K Century will require a global redesign of industry and settlement. Houses must be removed from coasts and floodplains and rebuilt. Buildings will have to be redesigned to survive stronger winds. Brick apartment buildings with black tar roofs become solar ovens in the summer: They will have to be torn down and replaced.

"Climate shocks to the world's most important cities—such as New York, Tokyo and London—could shake up economies worldwide," reports *Scientific American* writer Kathryn S. Brown. "By 2090, increasing storm surges could dunk lower Manhattan under water every few years, flooding the World Trade Center and other financial district skyscrapers."

It's the Ecology, Stupid!

The W2K threat did not register with the presidential contenders, although [former] President Clinton acknowledged the problem in a remarkable and underreported speech in New Zealand. "Unless we change course," Clinton declared, "The seas will rise so high they will swallow whole islands."

Clinton told his New Zealand audience that it was "no longer necessary to burn up the atmosphere to build economies on oil and coal." The president also admitted that the US was the country most responsible for producing climate-changing gases.

Instead of pouring billions of dollars into an indefensible Son-of-Star-Wars missile system to defend the country from space, the US needs to invest in creating environmental bulwarks on the ground.

During the Cold War, the US built bomb shelters and stocked them with water and survival rations. Faced with a climate-wide Hot-and-Cold-War, government emergency planners should be building storm-shelters in tornado belts and hurricane zones and making sure that they are stocked with clean water, medicine, and foodstuffs.

The US desperately needs a government whose policies are guided by biologists and independent scientists, not a corporate welfare state orchestrated by Big Business lobbyists.

Faced with the specter of the Y2K computer glitch, the world's leaders demonstrated their ability to marshal vast amounts of money and energy to forestall a threat to banking and commercial interests.

A similar effort to re-engineer the world economy to run on renewable solar power (wind, sun and wave) is possible and long overdue. It is the only way that the world can hope to survive the W2K Century. It is time to close the book on the dying carbon economies of the 20th century.

"The modest global warming now predicted should bring back one of the most pleasant and productive environments humans— and wildlife—have ever enjoyed."

The Effects of Global Warming Will Be Beneficial

Dennis T. Avery

In the following viewpoint, Dennis T. Avery argues that the anticipated global warming of 3.5 degrees Fahrenheit over the twenty-first century will have many positive impacts. According to Avery a warmer earth will result in lush forests, a decrease in climate-related disasters, increased food production, and a healthier human population. He describes a medieval period when temperatures increased by an amount similar to what is currently forecasted, and he concludes that this warming had a favorable impact on the world's population at the time. Avery is director of the Center for Global Food Issues within the Hudson Institute, a public policy research organization in Indianapolis.

As you read, consider the following questions:
1. According to the author, when did the earth's temperature warm by some 4 to 7 degrees Fahrenheit?
2. What effect will increasing atmospheric carbon dioxide have on plants and trees, in Avery's opinion?
3. In the author's opinion, what effect do equator-to-pole temperature gradients have on the weather?

Reprinted, with permission, from "Global Warming—Boon for Mankind?" by Dennis T. Avery, *American Outlook*, Spring 1998.

We've all read the global warming scare stories. . . . Fortunately, there is no reason for alarm. . . .

Climate researchers still do not agree on whether the earth will become warmer during the coming century. Even more importantly, none of them expect the planet to get very much warmer in the foreseeable future. They say that the earth is likely to warm by no more than 2 degrees Centigrade (3.5 degrees Fahrenheit) during the next century.

All the climate circulation models have cut their original warming forecasts at least in half in recent years, after satellite studies indicated that additional cloud cover would moderate any warming trend. Highly accurate satellite data for the last nineteen years show a slight cooling of the atmosphere. Most of the one-half-degree Centigrade of warming that has occurred in the last one-hundred years took place before 1940—before humanity put very much CO_2 into the air. Thus there is strong evidence that the two are unconnected.

Research has only recently produced a computerized climate model able accurately to mimic the weather the world has actually had. This more-accurate model projects only a 2 degree Centigrade increase in temperatures for the next century (see graph).

Medieval Global Warming

That may sound like a lot, but it isn't. The world has experienced that much warming, and fairly recently in history. And we loved it!

Between 900 AD and 1300 AD, the earth warmed by some 4 to 7 degrees Fahrenheit—almost exactly what the models now predict for the twenty-first century. History books call it the Little Climate Optimum. Written and oral history tells us that the warming created one of the most favorable periods in human history. Crops were plentiful, death rates diminished, and trade and industry expanded—while art and architecture flourished.

The world's population experienced far less hunger. Food production surged because winters were milder and growing seasons longer. Key growing regions had fewer floods and droughts. Human death rates declined, partly because of the decrease in hunger and partly because people spent less of

Global Warming Projections

°Centigrade warming projections for the year 2100

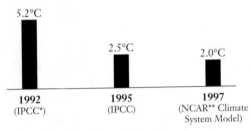

1992 (IPCC*)	1995 (IPCC)	1997 (NCAR** Climate System Model)
5.2°C	2.5°C	2.0°C

*Intergovernmental Panel on Climate Change
**National Center for Atmospheric Research

their time huddled in damp, smoke-filled hovels that encouraged the growth and spread of tuberculosis and other infectious diseases.

Prosperity, fostered by the abundant crops and lower death rates, stimulated a huge outpouring of human creativity—in engineering, trade, architecture, religion, art, and practical invention.

Soon after the year 1400, however, the good weather ended. The world dropped into the Little Ice Age, with harsher cold, fiercer storms, severe droughts, more crop failures, and more famines. According to climate historian H. H. Lamb, during this period, "for much of the [European] continent, the poor were reduced to eating dogs, cats, and even children." The cold persisted until the 18th century.

The Little Climate Optimum was a boon for mankind and the environment alike. The Vikings discovered and settled Greenland around 950 AD. Greenland was then so warm that thousands of colonists supported themselves by pasturing cattle on what is now frozen tundra. During this great global warming, Europe built the looming castles and soaring cathedrals that even today stun tourists with their size, beauty, and engineering excellence. These colossal buildings required the investment of millions of man-hours—which could be spared from farming because of the higher crop yields. . . .

We know less about what went on in North America. We do know that the Great Plains, the upper Mississippi Valley, and the Southwest apparently received more rainfall than they do now. The Anasazi civilization of the Southwest grew abundant irrigated crops—and then vanished when the Little Optimum ended and the rainfall declined. The Toltecs and Aztecs built marvelous civilizations in Mexican highlands that were plentifully watered.

Thus, we can cast aside the forecasts that global warming will bring more drought and expanding deserts. Global warming brings more clouds and more rainfall, especially near the equator. That is what apparently happened during the Little Optimum. For instance, North Africa received more rain than today, and the Sahara—and presumably many other desert regions—shrank in response to the increase in rainfall.

There were some negatives, of course. The steppes of Asia and parts of California apparently suffered dry periods. Also, it is important to remember that today's climate models are not precise enough to tell us anything about local rainfall in the future. The British global circulation model recently predicted that the Sahara Desert and Ireland would get exactly the same rainfall in the twenty-first century. That certainly seems unlikely.

Agricultural Bonanza

The medieval experience with global warming should reassure us greatly, and the latest scientific evidence supports such optimism. It is clear, for example, that a planet earth with longer growing seasons, more rainfall, and higher carbon dioxide (CO_2) levels would be a "plant heaven." Modest warming would help crops, not hinder them. There is virtually no place on earth too hot or humid to grow rice, cassava, sweet potatoes, or plantains, for example, and corn can be grown in a wider variety of climates than any other crop.

The prospective global warming will not be uniform. It is expected to moderate nighttime and winter low temperatures more than it raises daytime and summertime highs. Thus, it will produce relatively little added stress on crop plants or trees—and on people.

The expected increase in CO_2 will be an additional blessing. Carbon dioxide acts like fertilizer for plants. Dutch greenhouses, for example, routinely and deliberately triple their CO_2 levels—and the crops respond with 20 to 40 percent yield increases. Extra CO_2 also helps plants use their water more efficiently. The "pores" (stomata) on plant leaves partially close, and less water vapor escapes from inside the plants. More than a thousand experiments with 475 crop plant varieties in 29 separate countries show that doubling the world's carbon dioxide would raise crop yields an average of 52 percent.

The amount of carbon dioxide in the atmosphere does seem to be rising. In fact, we are nearly halfway to the expected CO_2 peak of 550 parts per million. The current levels of CO_2 in the earth's atmosphere are very low, however, compared to past periods. In fact, most of the earth's species of plants and animals evolved in much-higher levels of carbon dioxide than we have today—up to twenty times the recent pre-industrial level of 280 ppm.

Lush Forests and Prairies

The increase in CO_2 will make forests all over the world healthier and more robust—and allow them to support more wildlife. Canadian forestry researchers estimate that in a new warming their forest growth would increase by 20 percent. In fact, the world's crops, forests, and soils may well be nature's "missing carbon sink." (Not all human-produced carbon dioxide shows up in the atmosphere or is absorbed by the surface layers of the ocean, which suggests that it is being used by plants.)

Of course, it would put less stress on our wild species if the world always stayed at the same temperature, but the planet has never done that. Our "species models" mostly evolved in the Cambrian Period (six-hundred-million years ago), and they have already survived several Ice Ages and hot spells.

Scientists examining the impact of global warming on wildlife species in the two most at-risk environments (tropical forests and the Arctic) say that they would expect a modest global warming to produce little or no species loss.

In *Global Warming and Biodiversity*, for example, Dr. Gary S. Hartshorn notes that the tropical forests already undergo enormous variability in rainfall. He writes, "It is unlikely that higher temperature per se will be directly deleterious to tropical forest [wildlife] communities." Hartshorn also notes that although scientists previously estimated the number of wildlife species in the world at three to ten million, they had to change their estimate once they started counting tropical species. Now they estimate that there are roughly thirty million species, with the overwhelming majority occupying the tropical rain forests. Thus, the negligible effect of global warming on tropical forests bodes very well for the world's biodiversity.

In the same book, Dr. Vera Alexander notes that Arctic marine systems would be seriously threatened if the sea ice melted. The Arctic, however, has already survived major temperature changes, including the Little Climate Optimum, without shrinking appreciably. Even with average worldwide temperatures six to nine degrees Centigrade warmer than today's, Alexander notes, the sea ice would re-form in the winter.

Assessing an Arctic tundra ecosystem, Dwight Billings and Kim Moreau Peterson predict that such a warming would have no major species impact. They expect more snow-free days in the summer, more photosynthesis, and somewhat more peat decomposition, but these factors would mainly benefit the primary food chain. Thus the available evidence suggests that global warming will have little effect on Arctic species.

Of course, we must also note that any wildlife species too fragile to survive this kind of mild warming probably disappeared from the planet several hundred years ago during the Little Climate Optimum.

Decrease in Disasters

Most of the trillion-dollar estimates of global warming "costs" headlined in the 1980s were based on forecasts that cities such as New York City and Bangladesh would be drowned under rising seas. In 1980, for example, some activists claimed that global warming would raise sea levels by twenty-five feet. In 1985, a National Research Council panel estimated a three-foot rise in the sea level. Those are frightening scenarios, but completely untrue.

The Medieval Climate Optimum did not produce devastating floods. Nor will a new global warming. It may seem paradoxical, but a modest warming in the polar regions will actually mean more arctic ice, not less. The polar ice caps depend on snowfall, and polar air is normally very cold and dry. If polar temperatures warm a few degrees, there will be more moisture in the air—and more snowfall, and more polar ice.

The world's ocean levels have been rising at approximately the same rate—7 inches per century—for at least a thousand years. No one knows why. Data from the warming of 1900 to 1940 show a drop in sea levels and then a sea-level rise during the subsequent cooler period. In 1992, *Science* magazine published a paper based on ice core studies suggesting that the projected warming would *reduce* the sea level by one foot.

Global warming scaremongers have also claimed that a warmer world would suffer more extreme weather events. This too is unlikely. History records that the Little Optimum brought *fewer* floods and droughts. There is good reason to believe that this pattern would repeat in a new Little Optimum. Dr. Fred Singer, professor emeritus of Environmental Sciences at the University of Virginia, says, "One would expect severe weather to be less frequent because of reduced equator-to-pole temperature gradients."

In other words, the smaller the temperature difference between the North Pole and the equator, the milder the weather. Most of the warming, if it occurs, will be toward the poles, with very little increase near the equator. Thus, there would be less of the temperature difference that drives big storms.

Forging onward intrepidly, some alarmists have claimed that a warmer world would suffer huge increases in deaths from horrible plagues of malaria, yellow fever, and other warm-climate diseases. One study predicted fifty- to eighty-million more cases of malaria alone per year. (There are now approximately five-hundred million new cases of malaria each year, and up to 2.7 million deaths.)

Fortunately, these claims are unlikely to come true, because they ignore some important, fundamental realities. As mentioned, global warming would be very slight near the equator and thus would only slightly expand the range of the malarial mosquitoes. Hence there is little reason to expect tropical plagues to increase naturally.

Moreover, these diseases are nowhere near as relentless as the scare scenarios assume. In the U.S., for example, malaria and yellow fever once ranged from New Orleans to Chicago. We conquered those diseases, however, and not by changing the climate. We did it by suppressing mosquitoes, creating vaccines, and putting screens on doors, windows, and porches. Other countries can do the same. Third World countries have had high disease rates because they were poor, not because warm climates cannot be made safe.

As it happens, far from creating a plague of pestilences, the Little Climate Optimum engendered a worldwide population surge and set the stage for several historic invasions such as the Viking incursions into Normandy and England and the movement of German peoples into Eastern Europe. This time, however, global warming is quite unlikely to produce a population surge. The modern world's population is currently restabilizing thanks to affluence, urbanization, and contraceptive technology. Births per woman in the Third World have fallen from 6.5 in 1960 to 3.1 today. The First World is already below the replacement level (2.1 births) and likely to stabilize at the modern equilibrium of about 1.7 births per woman.

Warming or no, we can expect a peak population of approximately 8.5 billion people around 2035. That peak will be followed by a slow, gradual decline through the rest of the 21st century.

Why Be Wary?

The original global warming scare stories were authored by eco-activists who have subsequently admitted that they were looking for ways to persuade people to live leaner lifestyles. To frighten us into lowering our living standards, they have announced a whole series of terrifying claims, most of which have already been proven wrong:

• *The Population Explosion.* Activists frequently warned us that the human population would reach fifteen billion, or fifty billion, or whatever astronomic level would collapse the ecosystem. We now know that affluence and contraceptives will give the world a peak population of 8.5 billion around the year 2035, followed by a slow decline in the late twenty-first century.

• *Acid Rain*. Activists warned us that acid rain from industrial pollution would destroy the forests in the First World. A billion dollars worth of research has shown that acid rain is a very minor problem due mainly to natural factors.

• *Cancer from Pesticides*. We are still looking for the first case of human cancer from pesticide residues, and the National Research Council says that we will probably never find one. Moreover, as the National Research Council reports, "A sound recommendation for cancer prevention is to increase fruit and vegetable intake." Thus pesticides are actually helping cut cancer rates by producing more plentiful, affordable, and attractive fruits and vegetables.

There is no reason to believe that the authors of the global warming scares have any special knowledge about the future climate. In fact, their leading scientist, Dr. Stephen Schneider, was predicting global *cooling* just a few years ago, and he candidly states that he is willing to misrepresent the facts if it will stir up the public over the "correct" causes. New climate models make it clear that he is wrong.

"But what if we're right?" the activists respond. History says that they are not. And the problem is, the "solutions" these activists recommend, however well intended, would leave much of the world without an energy system—and that would be deadly for both people and animals. If we were to triple the cost of coal, double the cost of oil, ban nuclear power, and tear out hydroelectric dams—which would be the result of the activists' approach—humanity would essentially be left without energy.

Solar and wind power are extremely expensive and undependable. Burning large amounts of renewable wood would destroy huge tracts of forest—and the animals that live there. And in a world of expensive energy, people would not be able to afford the window screens, latrines, clean water, and refrigeration that prevent millions of deaths per year. Diarrhea, due mainly to spoiled food and untreated water, is the number one child-killer on the planet. Refrigeration has helped cut stomach cancer rates by three-fourths in the First World.

The widespread poverty caused by expensive energy would reverse the current worldwide trend toward greater affluence, decreasing birth rates, and better health. The low-energy option would destroy millions of square miles of wildlife habitat. High energy taxes would all but destroy modern agriculture, with its tractors and nitrogen fertilizer (produced mainly with natural gas). Shifting back to draft animals would mean clearing millions of additional acres of forest to feed the beasts of burden.

Giving up nitrogen fertilizer would mean clearing five to six million square miles of forest to grow clover and other nitrogen-fixing "green manure" crops. The losses of wilderness would nearly equal the combined land area of the United States and Brazil.

History and the emerging science of climatology tell us that we need not fear a return of the Little Climate Optimum. If there is any global warming in the twenty-first century, it will produce the kind of milder, more-pleasant weather that marked the Medieval Little Optimum—with the added benefit of more carbon dioxide in the atmosphere and therefore a more luxuriant natural environment.

The modest global warming now predicted should bring back one of the most pleasant and productive environments humans—and wildlife—have ever enjoyed. We have nothing to fear but the fear-mongers themselves.

*"The . . . health effects of climate change . . .
are likely to be severe, and many, many
people across the world will die as a result."*

Global Warming Will Severely Harm Human Health

Paul Kingsnorth

Paul Kingsnorth, a journalist and deputy editor of *Ecologist* magazine, contends in the following viewpoint that global warming will cause severe human health problems and lead to the deaths of many people. According to the author, floods, heat waves, and hurricanes due to global warming are already affecting human health, injuring or killing thousands and leaving survivors to endure famine and malnutrition. Another threat to health posed by global warming is the increase in diseases spread by pests and insects. According to Kingsnorth, many disease-carrying insects will find more breeding grounds as the world warms, spreading malaria and cholera to Europe and North America.

As you read, consider the following questions:

1. What percentage of the African population is expected to suffer from hunger and malnutrition in the future as a result of global warming, according to the Hadley Centre?
2. In the author's opinion, what caused the emergence of the Hantavirus Pulmonary Syndrome in the United States?
3. According to Kingsnorth, where has global warming enabled malaria to spread recently?

Reprinted, with permission, from "Human Health on the Line," by Paul Kingsnorth, *The Ecologist*, vol. 29, no. 2, March/April 1999, p. 92.

"Climate Change is likely to have wide-ranging and mostly adverse impacts on human health, with significant loss of life." Thus, the 2,000 scientists of the Intergovernmental Panel on Climate Change (IPCC) condemn future generations to the sort of battle against deadly diseases which the wonder drugs and scientific miracles of the twentieth century were supposed to have banished forever.

Of the many scientists who have projected, predicted and warned of the likely health effects of climate change, almost all agree on the basics: they will be widespread and unpredictable, they are likely to be severe, and many, many people across the world will die as a result.

The likely health effects are best divided into two categories: direct and indirect. Direct effects will result from direct exposure to the weather extremes that climate change will cause, for example: heat-stroke, hypothermia and deaths or injuries resulting from tidal waves, floods, hurricanes etc. Indirect effects will result from subsequent changes in environment and ecosystems—for example: the spread of vector-borne diseases into new areas, nutrition problems resulting from crop failure, diseases spread by algal blooms in warming seas, and even the mental health problems which may result from social and political dislocation.

Direct Effects of Global Warming

We are already seeing examples of some of the more obvious direct effects of climate change on human health. Just a few examples give some idea of the scale of the problem. In 1996, North Koreans were reduced to eating leaves and grass, following flash floods that destroyed their crops. Many suffered from malnutrition. That same year, 60 people in Spain died after a flash flood in the Pyrenees. In 1997, the worst rains in 30 years destroyed half of all Bolivia's crops, with hunger resulting, and a November typhoon in Vietnam resulted in 2,500 dead or missing. In 1998, heat-waves in India and the mid-USA killed over 4,000 people. Hurricane Mitch in Central America killed or injured an estimated 11,000. The Indonesian forest fires, started by man and exacerbated by warmer and drier-than average-weather, caused a massive increase in respiratory illnesses; crops were drowned in several countries, and fisheries failed, leading to an increase in hunger. Almost every one of these events was record-setting or breaking, and there are hundreds more such examples that could be quoted.

Predictions for the future point to more—much more—of the same, on a wider scale. The latest predictions from the UK's Hadley Centre, published in 1998 and based on an up-dated computer model, predict that at least 170 million people will be living in areas which are "extremely stressed" through lack of water in the twenty-first century, with death and severe illness the likely result. The Hadley Centre also predicts that 18 per cent more of the African population will suffer from hunger and malnutrition due to climate change than at present, and that, globally, over 20 million extra people each year will be at risk of flooding.

One health problem that is likely to become much more widespread in the 21st century is that known by scientists as 'thermal stress'—in everyday language, the effects of getting too hot or too cold, particularly during ever-more-frequent heat-waves and extreme winters. Detailed studies of the ef-fects of extreme weather on mortality rates have been con-ducted in many countries and, unsurprisingly, report a close correlation, particularly amongst children, the elderly and the infirm. Deaths from stroke, various cardiovascular ill-nesses, heat-stroke, hypothermia and influenza, in particular, are much more common during extremes of weather: and this applies to 'developed' as well as 'developing' countries.

Indirect Effects of Global Warming

A recent issue of *New Scientist* magazine reported that "hu-man disease is emerging as one of the most sensitive, and dis-tressing, indicators of climate change." It is accepted by vir-tually all climate scientists that the likely increase in, and spread of, potentially fatal diseases is likely to be the single most dangerous threat that climate change poses to human health. If many of the direr predictions are right, the flower-ing of diseases as the climate changes is very likely to negate the benefits of twentieth-century medical advances, and see the rebirth of diseases currently assumed to be 'conquered'.

A major threat will come from an increase of so-called

'vector-borne' diseases—those spread by pests, insects and other small creatures, such as snails. Dr Paul Epstein, of the Harvard Medical School, who has produced numerous studies and reports on this subject, believes that the spread of these diseases could be even more serious than currently feared.

Unhealthy Ecosystems and Disease

'Pests' such as rodents, insects and weeds, points out Epstein, are 'opportunists': they reproduce rapidly, have huge broods and wide appetites, and can quickly overrun an ecosystem if left to themselves. In a healthy ecosystem, there will be enough predators—lizards, birds and bats to eat the mosquitoes; owls and snakes to eat the rodents, etc.—to keep pest populations under control. However, the effect of global climate change will be to destabilise ecosystems across the globe, and disrupt predator-prey relationships. The result, in many places, is likely to be a vast increase in disease-carrying pests.

Epstein recently produced a study showing that, in many parts of the world, climatic disruption is already causing rodent-borne diseases to spread—and in some cases actually causing new diseases to emerge. In the early 1990s in the USA, for example, a combination of prolonged drought—which killed predators such as coyotes, snakes and owls—followed by heavy rains, precipitated a ten-fold increase in the rodent population (rodents thrive in and around water, even if it is contaminated). This plague of rodents led to the emergence of a new disease—Hantavirus Pulmonary Syndrome—which was apparently transmitted to humans via the rodents' droppings. Similar hantaviruses have also emerged, in similar climatic conditions, in several European nations, particularly Yugoslavia, while other rodent-borne diseases, like leptospirosis and vital haemorrhagic fevers, have spread across Latin America.

Rising Temperatures and the Spread of Malaria

While localised and regional climatic changes are likely to lead to an increase in vector-borne diseases, the average global rise in temperature will also exacerbate the same trend. Many disease-carrying insects—most obviously the malarial mosquito—thrive in warm conditions; as the world warms, they will begin to find more places in which they can breed. A 1996 report from the London School of Hygiene and Tropical Medicine illustrated this point clearly when it calculated that, of ten of the world's most dangerous vector-borne diseases (malaria, schistosomiasis, dengue fever, lymphatic filariasis, sleeping sickness, Guinea worm, leishmaniasis, river blindness, Chagas' disease and yellow fever), all but one were likely to increase, or in some way change their range as a result of climate change.

Excessive Heat Will Increase Mortality

The main direct effect of climate change on humans themselves will be that of heat stress in the extreme high temperatures that will become more frequent and more widespread. Studies using data from large cities where heat waves commonly occur show death rates which can be doubled or tripled during days of unusually high temperatures. Although such an episode may be followed by a period when fewer deaths occur, showing that some of the deaths would in any case have occurred about that time, most of the increased mortality seems to be directly associated with the excessive temperatures. It might be thought that compensation for the periods of excessive heat would be provided by fewer occasions of severe cold. Studies tend to show, however, that increased mortality due to periods of excessive heat will considerably exceed any reduction in the periods.

John Houghton, *Global Warming*, 1994.

Malaria is the world's most prevalent mosquito-borne disease: two million people die from it every year. But it is likely to get worse. The scientist R. Colwell has called malaria "an old disease with the potential of re-emerging as a new disease, especially in association with climate change," and virtually all experts seem to agree that one effect of climate change will be to increase the range of the malarial mosquito. The IPCC predicts that malaria will spread from affecting 45 per cent of the population, as it does today, to affecting 60 per cent by the latter half of the 21st century—of the order of 50–80 million additional annual cases. The Hadley Centre's 1998 study predicted a significant spread in the mosquito's range, largely as a result of the warming of previously temperate areas—including parts of Europe and North America.

Malaria is also likely to spread to high altitude areas, such as the Andes, as their average temperature rises.

Again, it seems that in some places this is already beginning: malaria has already begun to affect the previously mosquito-free African highlands, and upland rural areas of Papua New Guinea. Urban centres are beginning to suffer as well: many central African cities are experiencing urban malaria for the first time, and two recent cases in New York City were traced to local mosquitoes. Furthermore Paul Epstein, in studying cases of malaria linked to the recent El Nino, has found that large and deadly outbreaks across Asia were one result of climatic upheavals there.

Other vector-borne diseases are likely to become more common—and hence more deadly—too. Again, the spread is already beginning. Mosquito-borne yellow fever has recently invaded Ethiopia, and dengue fever, spreading through the Americas, has already reached Texas. Recent floods in northeast Kenya caused Rift Valley Fever, a cattle disease, to jump the species barrier and kill hundreds of people. In 1994, the pneumonic plague resurfaced in India, during a summer in which temperatures reached as high as 124 degrees Fahrenheit. We should expect more of the same as the thermometers of many nations are thrown out of kilter by man-made climate change.

Less Clean Water, More Epidemics

And it is not just vector-borne diseases that are likely to take advantage of the changing climate. Other infectious killers are likely to enjoy a resurgence too, particularly diseases associated with water supply and sanitation. A 1996 World Health Organization (WHO) report laid out the threat starkly: "climate change could have a major impact on water resources and sanitation by reducing water supply. This could in turn reduce the water available for drinking and washing, and lower the efficiency of local sewerage systems, leading to increased concentration of pathogenic organisms in raw water supplies."

This was the situation in 1991, when Peru was devastated by a cholera epidemic (which quickly spread across Latin America, killing over 5,000 people in eighteen months) linked with the warmer waters of El Nino. "Of course," wrote Karen Schmidt, in the *New Scientist*, "it is really global warming that is involved." Cholera, often assumed to be largely a disease of the past, may well become common again in the 21st century, as global warming bites. Paul Epstein's pioneering work has also pointed out a hidden threat in this area, too: not only is cholera associated with poor sanitation and polluted inland waters, but it can also be harboured in marine plankton. Epstein believes that this was the original cause of the 1991 epidemic in Latin America.

Apart from cholera, other water-borne and water-related-diseases are also likely to increase and spread too, for the same reasons: typhoid, hepatitis A, diarrhoeal diseases (major killers of young children in 'developing' countries), scabies, trachoma and schistosomiasis, to name but a few. But water-and-climate-change-caused diseases are linked in another way, too: the ocean itself could become, and may even already be becoming, a new vector for fatal diseases.

Warming Oceans Breed Disease

In January 1999, the *New York Times* reported that previously unknown bacteria, fungi and viruses are beginning to bloom in the oceans as they warm, killing coral and fish, and threatening human health. Joan B. Rose, from the University of South Florida, reported that human viruses were spreading into the warming seas from the 1.8 million septic tanks along the Florida coast. "Many people are becoming infected with viruses picked up while swimming, windsurfing or bathing in infected waters," she confirmed. James W. Porter, from the University of Georgia, believes that this unprecedented problem is linked to a rise of 1.8 degrees Celsius in ocean temperature which climate change has already caused in the area.

Paul Epstein has studied the relationship between climate change, ocean pollution and disease, too, and has produced equally worrying conclusions. His suggestion that cholera can be transmitted by marine plankton has already been mentioned, but he has also postulated that coastal algal blooms already being seen in many of the world's seas—as a direct result of the warming of the water—are also transmitters of disease, often via fish and shellfish. In the summer of 1992, for example, after a long warm period, blooms known as Alexandrium tamarense developed in the seas around Newfoundland, infecting shellfish with a disease known as Paralytic Shellfish Poisoning (PSP) which was transmitted to humans who ate the shellfish. Similar PSP incidents have since occurred in waters around the east coast of the USA, Canada and the UK. Toxic 'brown tides' have poisoned scallops and eels, and numerous other episodes of fish intended for human consumption being poisoned by algal blooms have been catalogued by Epstein.

Costly Inaction

One obvious, but often overlooked, consequence of the health problems which climate change is preparing to visit on us, is the financial cost of dealing with the problem. Economists and industrialists who insist that taking any action to combat climate change will threaten the world's economies might like to consider the economic costs of doing nothing. In terms of human health, some of those costs are already being borne: Epstein reports that the 1991 cholera epidemic cost Peru over $1 billion, while airline and hotel industries lost between $2 billion and $5 billion from the 1994 Indian plague. Cruise boats are already avoiding islands in the Indian Ocean plagued by dengue fever—and are threatening the area's $12 billion tourist industry in the process.

This viewpoint has only scratched the surface of this issue. It has not even begun to address some of the threats that are more difficult to predict, such as the potential diseases of the mind which could stem from the chaos caused by a changing climate what psychiatrists call the 'psychosocial' problems associated with economic collapse, institutional breakup and social upheaval. Hundreds of papers, millions of words and many laboratories and books have been dedicated to predicting the likely effects of climate change on human health. But the simple fact is that many of those effects are likely to catch us unawares. In medical terms, it is more than likely that, as Paul Epstein succinctly puts it, we are "vastly underestimating the true costs of 'business-as-usual'; and underestimating the benefits to society as a whole of using the resources we have inherited efficiently."

"*A warmer climate should improve health and extend life . . . [and] is likely to prove positive for human health.*"

Global Warming Will Improve Human Health

Thomas Gale Moore

According to Thomas Gale Moore in the following viewpoint, global warming will not lead to an increase in infectious disease and will improve human health by lowering the number of deaths due to weather-related cold. In the author's opinion, the argument that global warming is responsible for the spread of mosquito-borne diseases from the tropics to temperate latitudes has little basis in reality. He claims that most new cases of malaria, cholera, and yellow and dengue fevers in the United States can be attributed to immigration. In fact, according to Moore, these diseases caused many more deaths in the United States during the statistically colder nineteenth and early twentieth centuries. Moore is a senior fellow at the Hoover Institution, an organization dedicated to limiting government intrusion into the lives of individuals.

As you read, consider the following questions:
1. In the author's opinion, what explains the absence of heat-related deaths in southern cities?
2. On average, how many cases of malaria does Moore claim are reported annually in the United States?
3. According to Moore, what explains the outbreak of cholera in Peru in 1991?

Reprinted from *Climate of Fear: Why We Shouldn't Worry About Global Warming*, by Thomas Gale Moore (Washington, DC: Cato Institute, 1998), by permission of the publisher.

M any researchers, environmentalists, and politicians forecast that rising world temperatures in the twenty-first century will have devastating effects on human health. Referring to the world as a whole, Working Group II of the Intergovernmental Panel on Climate Change (IPPC) asserted: "Climate change is likely to have wide-ranging and mostly adverse impacts on human health, with significant loss of life." The authors of the IPCC report feared that increases in heat waves would cause a rise in deaths from cardio-respiratory complications. They also foresaw a rise in vector-borne diseases, such as malaria and dengue and yellow fevers. The report did acknowledge briefly that, in colder regions, there would be fewer cold-related deaths.

Public Health and Global Warming

Most of the causes of premature death have nothing to do with climate. Worldwide the leading causes are chronic diseases—accounting for 24 million deaths in 1996—such as maladies of the circulatory system, cancers, mental disorders, chronic respiratory conditions, and musculoskeletal disorders, none of which has anything to do with climate but everything to do with aging. Another 17 million, most of them in poor countries, succumbed in the same year to disorders caused by infections or parasites, such as diarrhea, tuberculosis, measles, and malaria. Many of those diseases are unrelated to climate; most have to do with poverty.

Diarrheal diseases, such as cholera and dysentery, killed 2.5 million of the 52 million people who died worldwide in 1996. Through the provision of fresh water and proper sanitation, those diseases are easily preventable. Although a warmer climate might make the environment more hospitable for such afflictions as cholera, dysentery, and typhoid in areas without good sanitation or clean water, chlorination and filtration could halt their spread.

Both the scientific community and the medical establishment assert that the frightful forecasts of an upsurge in disease and early mortality stemming from climate change are unfounded, exaggerated, or misleading and do not require action to reduce greenhouse gas emissions. *Science* magazine reported that "predictions that global warming will spark epidemics have little basis, say infectious disease specialists, who argue that public health measures will inevitably out-weigh effects of climate." It added: "Many of the researchers behind the dire predictions concede that the scenarios are speculative.". . .

The following examines the effect of climate and, in particular, temperatures on mortality in the United States. Anecdotal evidence suggests that warmer temperatures may actually promote health. Folklore alleges that physicians sometimes recommend that patients escape to a more clement climate, never to a colder one. . . .

Hot Weather and Death Rates

Recent summers have sizzled. Newspapers have reported the tragic deaths of the poor and the aged on days when the mercury reached torrid levels. Prophets of doom forecast that rising temperatures in the twenty-first century portend a future of calamitous mortality. Scenes of men, women, and children collapsing on hot streets haunt our imaginations.

Happily the evidence refutes that dire scenario. First, however, let us review the documentation supporting the supposition that human mortality will rise with rising temperatures. Death rates during periods of very hot weather have jumped in certain cities, but above-normal mortality has not been recorded during all hot spells or in all cities. Moreover, research concerned with "killer" heat waves has generally ignored or downplayed the reduction in fatalities that warmer winter months would bring. . . .

Researchers analyzing hot days and deaths have found no constant relationship; even when extremes in weather and mortality are correlated, the relationship is inconsistent. Cities with the highest average number of summer deaths are found in the Midwest or Northeast while those with the low-est number are in the South. Typically analysts have failed to find any relationship between excess mortality and tempera-ture in southern cities, which experience the most heat. Other studies have found that people who move from a cold to a subtropical climate adjust within a very short period. . . . .

Researchers have attributed the absence of heat-related deaths in southern cities to acclimatization and the prevalence of housing that shields residents from high temperatures. In the North, the housing of the elderly and the poor is usually old and dilapidated. Over the next hundred years, if not sooner, most of those buildings will be torn down and replaced. Should the climate warm, builders will move toward structures that protect the inhabitants from extreme heat, as housing in the South allegedly does now. . . .

Heat-stress does increase mortality; but it typically affects only the old and infirm, whose lives may be shortened by a few days or perhaps a week. There is no evidence, however, that general mortality rises significantly. The numbers of heat-stress-related deaths are very small; in the United States they are exceeded by the number deaths due to weather-related cold. During the latest 10-year period for which we have data (see graph) which includes the very hot summer of 1988, the average number of weather-connected heat deaths was 132, compared with 385 for those who died from cold. Even during 1988, more than double the number of Americans died from the cold than passed on from the heat of summer. A somewhat warmer climate would clearly reduce more deaths in the winter than it would add in the summer.

Mosquito-Borne Diseases

A growing chorus has been chanting that global climate change will spread insect-borne diseases, such as malaria, dengue fever, and yellow fever, to temperate latitudes. In 1996, the health effects of global warming were the subject of lengthy journal articles in the *Journal of the American Medical Association* (1996), and *Lancet* (1996), an international journal of medical science and practice. In September 1996, the Australian Medical Association sponsored a major conference on the subject. Professor Paul Epstein of the Harvard School of Public Health claimed that in the past few years mosquitoes carrying malaria and dengue fever had been found at higher altitudes in Africa, Asia, and Latin America. In North America, David Danzig, in a Sierra Club publication, has contended that only the tip of Florida is currently warm enough to support malaria-carrying mosquitoes but that global warming could make most of us vulnerable. He should check his history.

Proportion of Weather-Related Cold Deaths to Heat-Stress Mortality

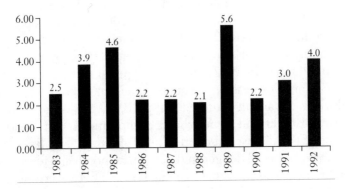

Vital Statistics of the United States (1983–1992).

Before the Second World War, malaria was widespread in the United States. The Centers for Disease Control and the *Statistical Abstract of the United States* for the relevant years reported that over 120,000 cases were reported in 1934; as late as 1940, the number of new sufferers totaled 78,000. After the war, reported malaria cases in the United States plunged from 63,000 in 1945 to a little over 2,000 in 1950 to only 522 in 1955. By 1960, DDT had almost eliminated the disease; only 72 cases were recorded in the whole country. In 1969 and 1970, the CDC reported a resurgence to around 3,000 cases annually, brought in by service personnel returning from Vietnam. Subsequently, immigrants from tropical areas have spawned small upticks in new cases.

In the 1980s and 1990s, the number of reported cases averaged around 1,200 to 1,300 annually. The CDC reports that since 1985 approximately 1,000 of those cases have been imported every year, with visitors and recent immigrants accounting for about half. The rest come from travelers arriving from tropical countries, service personnel returning from infested areas, and a handful of individuals, typically those living near international airports, bitten by a mosquito that hitched a ride from a poor country. More stringent efforts to keep out the unwanted "immigrants" may be called for if the problem worsens.

Yellow and dengue fevers were both common in the United States from the 17th century onward. Epidemics of yellow fever ravaged New Yorkers and killed tens of thousands of people. In one year, 1878, of 100,000 cases reported along the East Coast, 20,000 people died. Between 1827 and 1946, eight major pandemics of dengue fever overran the United States. In 1922, the disease spread from Texas, with half a million cases, through Louisiana, Georgia, and Florida. Savannah suffered with 30,000 cases, of which nearly 10,000 had hemorrhagic symptoms, a very serious form of the disease. In contrast, in 1996 the CDC listed 86 imported cases of dengue and dengue hemorrhagic fever and eight local transmissions, all in Texas. There were *no* reported cases of yellow fever.

As a public health issue, those diseases, which did plague the United States in the reputedly colder 19th and early 20th centuries, have been largely exterminated. There is no evidence that a resurgence is imminent. Certainly the climate is not keeping the spread of the diseases in check. If it was warm enough in the cold 19th century for the mosquitoes to thrive, it is warm enough now! . . .

Living Standards and the Spread of Disease

What brought an end to the scourges? The introduction of DDT clearly played a major role. From the end of World War II until it was banned in 1972, the pesticide worked wonders to eliminate harmful insects, especially mosquitoes. But it was not just insecticides that did the trick. Simple steps, such as screens on windows, the elimination of standing water, and the movement to the suburbs, which reduced population density and thus the risk of transmission, have played a critical role in eliminating mosquito-borne diseases.

In 1995, however, a dengue pandemic afflicted the Caribbean, Central America, and Mexico, generating around 74,000 cases. Over 4,000 Mexicans living in the Tamaulipas state, which borders Texas, came down with the disease. Yet Americans living a short distance away remained unaffected. The contrast between the twin cities of Reynosa, Mexico, which suffered 2,361 cases, and Hidalgo, Texas, just across the border, is striking. Including the border towns, Texas reported only 8 nonimported cases for the whole state.

The only reasonable explanation for the difference between the spread of dengue in Tamaulipas and its absence in Texas is living standards. Where people enjoy good sanitation and public education, have the knowledge and willingness to manage standing water around households, implement programs to control mosquitoes, and employ screens and air-conditioning, mosquito-borne diseases cannot spread. If the climate does warm, those factors will remain. In short, Americans need not fear an epidemic of tropical diseases.

Cholera

A recent manifestation of fear-mongering about the health effects of global warming is a curious article in *Science*, taken from a modified text of Rita Colwell's presidential address to the American Association for the Advancement of Science's 1996 annual meeting. This address presents a studious analysis of cholera and its recent resurgence in the Americas. What is most singular is not what is in Dr. Colwell's report but what she does not mention. . . .

First, a few dry facts about cholera, an infectious disease caused by *Vibrio cholerae*, a bacterium that can bring on diarrhea, vomiting, and leg cramps. Without treatment, a person can rapidly lose body fluids, become dehydrated, and go into shock. Death can come quickly. Treatment is simple, the replacement of the fluids and salts with an oral rehydration solution of sugar and salts mixed with water. Fewer than 1 percent of those who contract cholera and are treated die.

Cholera cannot be caught from others but comes from ingesting food or water that contains the bacterium. Eating tainted shellfish, raw or undercooked fish, raw vegetables, or unpeeled fruits can lead to infection. Drinking unpurified water can be dangerous as well. The bacterium thrives in brackish warm water but can survive, in a dormant state, both colder water and changes in salinity. *V. cholerae* is also associated with zooplankton, shellfish, and fish. It often colonizes copepods, minute marine crustaceans. Ocean currents and tidal movements can sweep the bacterium, riding on copepods, along coasts and up estuaries where *V. cholerae* can remain dormant until conditions are ripe for it to multiply. . . .

Before the most recent outbreak, the world suffered six cholera pandemics. By the end of the 19th century, however, Europe and North America were free of the disease. The solution was simple: filtration and chlorination of the water supply. Filtering alone not only reduces the spread of cholera but cuts typhoid significantly. Combining filtration with chlorination eliminates waterborne diseases. A warmer climate, if it were to occur, would not reduce the effectiveness of water purification measures.

Misplaced Environmentalism Causes Outbreak

In January 1991, after many disease-free decades, cholera began sickening villagers in Chancay, Peru, a port less than 40 miles north of Lima. It then spread rapidly up and down the coast. From that outbreak to the end of 1995, Latin America reported over 1 million cases—many went unreported—and 11,000 deaths. The illness traveled from Peru to Ecuador, Colombia, then Brazil. Eight months after appearing in Peru, it reached Bolivia. By the end of 1992, virtually all of South and Central America, from Mexico to Argentina, had confirmed cases. In the early 1990s, cholera also entered the United States; however, with the exception of a few cases brought on from eating raw, tainted shellfish, virtually all cases were contracted abroad. Seventy-five cases, nearly half the total 160 reported to the CDC between 1992 and 1994, originated on a single flight from Lima in 1992!

What went wrong to bring an end to Latin America's 100 years of freedom from cholera? Rita Colwell theorizes that an El Niño* led to a plankton bloom that multiplied the hosts of *V. cholerae*. But El Niños have been occurring with some regularity for many years without producing a cholera epidemic. The coast of Peru in 1991 was not even particularly warm compared with a number of other years. Even if El Niño were in part the culprit, the basic cause lies elsewhere. On the basis of Environmental Protection Agency (EPA) studies showing that chlorine might create a slight cancer risk, authorities in Peru decided not to chlorinate

* A warming of the ocean surface off the western coast of South America that occurs every 4 to 12 years when upwelling of cold, nutrient-rich water does not occur. It affects weather over much of the world.

their country's drinking water. Perhaps they also thought they would save money. Chlorination, however, is the single most effective preventive of cholera and other waterborne diseases. After the fiasco in Peru, the EPA determined in 1992 that there was no demonstrable link between chlorinated drinking water and cancer. It was too late; the harm had been done. Peru's misplaced environmentalism led to more than 300,000 victims in that country alone.

Cholera is a disease of poverty, crowding, and unsanitary conditions. A warmer climate will not carry the disease to affluent countries; but in the Third World, economic growth can bring freedom from it and many other diseases. We should not impose costs on ourselves or others that would reduce the resources needed to bring clean water and good sanitation to Latin America, Africa, and Asia. . . .

Seasonal Effects

If climate change were to manifest itself as warmer winters without much of an increase in temperature during the hot months, which some climate models predict, the change in weather could be especially beneficial to human health. The IPCC reports that, over the twentieth century, the weather in much of the world has been consistent with such a pattern: winter and night temperatures have risen while summer temperatures have fallen. . . .

According to the National Climatic Data Center, a sample of 45 metropolitan areas in the United States shows that for each increase of a degree in the average annual temperature, July's average temperatures go up by only 0.5°F while January's average temperatures climb by 1.5°F. Since warming will likely exert the maximum effect during the coldest periods but have much less effect during the hottest months, the climate change should reduce deaths even more than any summer increase might boost them.

Deaths in the United States and most other advanced countries in the middle latitudes are higher in the winter than in the summer. Except for accidents, suicides, and homicides, which are slightly higher in the summer, death rates from virtually all other major causes rise in winter months; overall mortality from 1985 to 1990 was 16 percent greater when it was cold than during the warm season. These data suggest that, rather than increasing mortality, warmer weather should reduce it; but that possibility is rarely discussed. . . .

Human Health Will Benefit

Although it is impossible to measure the gains exactly, a moderately warmer climate would be likely to benefit Americans in many ways, especially in health. At the same time, let me stress that the evidence presented here is for a *moderate* rise in temperatures. If warming were to continue well beyond 4.5°F, the costs would mount and at some point the health effects would undoubtedly turn negative. Contrary to many dire forecasts, however, the temperature increase predicted by the IPCC, which is now less than 4.5°F, under a doubling of greenhouse gases would yield health benefits for Americans.

In summary, a warmer climate should improve health and extend life, at least for Americans and probably for Europeans, the Japanese, and people living in high latitudes. High death rates in the tropics appear to be more a function of poverty than of climate. Thus global warming is likely to prove positive for human health.

"The threat of sea level rise spans an
enormous range of possible impacts. . . .
And all the possible repercussions are
assuredly negative."

Global Warming Will Cause Sea Levels to Rise

Stuart R. Gaffin

In the following viewpoint, Stuart R. Gaffin finds that rising sea levels as a result of global warming are inevitable and that understanding the effects of the rise will be crucial in taking preventive measures against flooding. According to Gaffin, forecasts suggest that during the twenty-first century, thermal expansion of warming oceans and the melting of glaciers and ice caps will cause a sea level rise of up to one meter. Small island nations are currently facing the greatest risk from sea level rise and will meet with increased erosion, greater energy demands, coral reef deterioration, loss of income, and destabilized human settlements, in the author's opinion. Gaffin is a staff scientist with the Environmental Defense Fund, an organization working to protect the environmental rights of all people.

As you read, consider the following questions:
1. How many people around the globe does Gaffin claim are living in the flood zone?
2. In the author's opinion, how far would beaches retreat along the California coast if sea levels were to rise by one meter?

The accumulation of anthropogenic greenhouse gases in the atmosphere, principally carbon dioxide but also methane and nitrous oxide, is expected to cause substantial warming of the Earth. Atmospheric concentration of carbon dioxide has increased 30% over its pre-industrial level. Without policy intervention, it will continue to increase to much higher levels over the next century due to increasing energy demand related to development and population growth. Climate models project that, as a result, the Earth's surface could warm anywhere between 2 and 6 degrees Fahrenheit (1 to 3.5 degrees Celsius). Many ecological consequences will ensue, among the most direct of which will be the melting of small glaciers and ice caps on land and the expansion of seawater as it warms. Both these effects will cause global sea level to rise. This viewpoint discusses the impacts of that rise on selected coastlines and islands.

Many case studies estimating land loss and other impacts of global sea level rise have been conducted around the world and reported in the scientific assessment reports of the Intergovernmental Panel on Climate Change (IPCC). By convention most of these studies have reported impacts associated with benchmark sea level rises of 1 or 0.5 meters. Obviously, these impacts will not occur as isolated "events" in the future, but rather as a continuing and accelerating process of impacts, and large land losses will take place with sea level increases below 1 meter and also below 0.5 meters.

This viewpoint surveys the literature of impacts and, using downward proportioning when necessary, illustrates a range of possible impacts below 1 meter sea level rise both for selected sites worldwide and for the United States. It also superimposes these impacts onto the IPCC projections for future global sea level rise. . . .

The threat of sea level rise spans an enormous range of possible impacts from the relatively small and manageable to the catastrophic. And all the possible repercussions are assuredly negative.

One must include in any calculation of the effects of sea level rise a rapidly growing human population that relies heavily on coastal lands for food, recreation and natural resources. The majority of the world's people live near sea level in large coastal cities or on coastal plains. Although more work needs to be done to quantify the number, a commonly cited estimate is that 50 to 70 percent of humanity lives within the coastal zone. More relevant and rigorous though is the estimate that 46 million people, mostly in developing countries, presently live in the flood zone and are exposed to a storm surge in an average year and that this number would double if sea level rises 50 centimeters (cm).

As with current population distribution, population *growth* over the next few decades will also be concentrated near the sea. However, like picnickers on the beach ignoring the coming high tide, we show little inclination to retreat from the edge. For this reason, some have referred to the trends of sea level rise and coastal population growth as a "collision course."

Responding to sea level rise by retreat, accommodation, or protection will impose a complex set of hard choices on society that will vary widely around the globe. The choices made will be dictated by many issues, including geography, technology, human resources, politics, cultural acceptance, and economic considerations.

Estimates of Current and Past Century Sea Level Rise

The seas have risen over the past century. Based on global tide gauge data, the rate of rise has averaged between 1 and 2.5 millimeters (mm) per year over the past 100 years, with a best estimate of 1.8 mm/yr. Therefore over the past century the oceans have risen between 10 and 25 cm, with a best estimate of 18 cm (= 7 inches).

The past century of sea level rise is roughly consistent with that expected from models of oceanic thermal expansion and analysis of the retreat of the world's mountain glaciers: seawater expansion and glacial meltwater seem to have contributed roughly equal amounts. However, the uncertainties involved in this consistency check are still large and is an area needing further study.

There is compelling geologic and archaeological data that the rate of sea level rise during the past 100 years represents a significant acceleration of that over the previous 2,000 years. Such studies constrain the rate of rise over the previous two millennia at 0.4 mm/yr. There is no strong evidence that sea level rise has accelerated during the past few decades, but this conclusion depends critically on a small number of long tide gauge records. Measurements made with satellite radar altimeters will improve significantly the observational database on sea level rise in the future.

Projecting Future Sea Level Rise

The Intergovernmental Panel on Climate Change (IPCC) has published forecasts of future sea level rise using a variety of assumptions about future global warming and the physical processes leading to sea level increases. The projections are shown in the graph. These projections are referred to in the IPCC reports as the "extreme" range of possible sea level increases due to the various uncertainties, as explained below.

The projections were made using "business-as-usual" (i.e., no-further-climate-policy) emissions scenarios published by IPCC in 1992 and referred to as the "IS92 a-e" series. These scenarios, in turn, were based on a range of assumptions about future population growth, economic development, resource availability, and technological changes. The IS92e scenario is the "high" emissions projection, while the IS92c scenario is the "low" projection. The IS92a scenario is a widely used "best" estimate.

The emissions scenarios drive models of sea level rise that take into account both the thermal expansion of seawater as it warms and the thinning of mountain glaciers on land. Critical model estimates include the extent to which both the climate will warm in response to a given increase in atmospheric greenhouse gases and the glaciers will melt as a result of this warming.

Despite the large range in the emissions scenarios (6–35 billion tonnes of carbon in the year 2100), this uncertainty results in only a small spread in the projections for sea level. By far, most of the range shown in the graph is due to the climate sensitivity and ice melt parameter assumptions. The forecasts suggest that over the next century thermal expansion will contribute about 60 percent to sea level rise, while the remaining 40 percent will result from glacial, ice cap, and Greenland ice-sheet melting.

Also shown in the graph, for comparison, is the sea level projection that results from a scenario of greenhouse gas emissions that stabilizes atmospheric carbon dioxide at 450 parts per million (ppm) in the future. Although greenhouse gases are stabilized in this projection, global warming continues because of lags in the climate system and the sea level rises accordingly. Environmentalists and ecologists favor 450 ppm as the stabilization target to facilitate adaptive responses by ecosystems. However, even this low-end target results in considerable sea level rise that continues for centuries past 2100.

An important point to bear in mind when reviewing the graph is that the IS92a estimate represents a future rate of sea level rise that is two to five times that experienced over the past century (1–2 mm/yr). Thus the rate of land loss and increase in storm surges experienced in the past will not be an adequate guide to the losses that will be suffered and the adjustments that will be required in the future. . . .

General Impacts of Global Sea Level Rise

In assessing the general impacts of a rising sea, it should be noted that the world's coastlines (like the world's climate) have enjoyed relative stability for thousands of years following the end of the last ice-age cycle which terminated 10,000 years ago. Thus most coastal landforms have had time to achieve relatively stationary configurations, although changes are continually taking place due to such short-lived disturbances as earthquakes and storm events.

It is easily appreciated that a global sea level rise will lead to increased marine submergence of low-lying coastal areas, generally referred to as inundation. High and low tides will advance landward accordingly. The new high and low tides will lead to increased erosion as nearshore waves break farther inland.

Erosion is an important effect amplifying inundation, thus leading to even greater land loss. The magnitude of the erosion accompanying a sea level rise is generally determined by an equation called the "Bruun Rule." A simple explanation is given by James G. Titus and coworkers which is repeated here. The visible part of a beach is generally much steeper than the underwater portion, which comprises the active "surf zone." While the extent of inundation is determined by the slope of land just above water, the total shoreline retreat is determined by the average slope of the entire beach profile. The slope of the whole beach profile is generally shallower than that of the above-water portion, leading to greater land losses.

As an example, for the United States, a 1 meter rise in sea level would cause beaches to retreat (erode plus inundate) 50 to 100 meters from New England to Maryland, 200 meters along the Carolina coast, 100 to 1000 meters along the Florida coast, and 200 to 400 meters along the California coast.

Effects of Sea Level Rise on Small Island Nations

Small island nations, many of which are only a few meters above sea level, may be facing annihilation due to the inevitability of significant sea level rise as shown in the graph. Thus they deserve special attention in any report on sea level rise impacts. Among the most vulnerable of these islands are the Marshall Islands, Kiribati, Tuvalu, Tonga, the Line Islands, Federated States of Micronesia, Cook Islands (in the Pacific Ocean); Antigua and Nevis (in the Caribbean Sea); the Maldives (in the Indian Ocean). . . .

Among the impacts that small islands will face due to sea level rise and climatic warming are (1) increased coastal erosion; (2) changes in aquifer volumes with increased saline intrusion; (3) greater demand for air conditioning and hence energy consumption and fossil fuel importation; (4) coral reef deterioration resulting from sea level rise accompanied by thermal stress; (5) social instability related to inter-island migration; (6) loss of income through negative impacts on tourist resort location; and (7) increased vulnerability of human settlements because of increasing size.

Policies and practices that will exacerbate these problems include (1) coral reef mining; (2) land reclamation; (3) construction of harbors, jetties, and breakwaters; (4) overutilization of acquifer resources; (5) lack of in-country data covering physical and biological resources; (6) shortage of manpower; and (7) inadequate disposal of sewage and toxic chemicals. The failure to adequately address current environmental problems will leave the islands more vulnerable to future climatic and sea level changes.

Climate change will affect rainfall, wind and monsoon patterns. Upwelling zones in the ocean may shift, affecting fisheries. Storms and droughts may increase. Evapo-transpiration rates may change. The general weather related changes may lead to impacts on agricultural crops, natural vegetation and the growing season.

Finally it is interesting to note that since marine resources provide the bulk of small island income, if they could continue to occupy the island through protection, they may remain economically viable. However protection measures will have to be carefully analyzed. Dikes, for example, are infeasible because of coral atoll porosity. Thus the islands would most likely have to elevate.

The complexity of the details of the impacts on small islands of a sea level rise is a useful prelude to keep in mind when considering the abbreviated land loss impacts shown in the graph.

Impacts of Sea Level Rise Superimposed on IPCC Projections

Abbreviated summaries of the consequences of sea level rise on selected coastlines and islands are presented in the graph, which focuses on global impacts. . . .

The graph superimposes captions for land loss impacts at various locations on top of the IPCC sea level rise projections. The impacts are keyed to "global/relative" sea level rises in 10-cm increments starting at 20-cm and increasing to 100-cm. . . .

The particular countries shown were selected because they are often cited for being especially vulnerable to sea level rise and are subject to large land-loss risks. . . .

The loss of land and the extent of vulnerability shown in the graph generally refer to losses that will be suffered without any attempts at protection, mitigation, or adaptation. Obviously, such human responses will be considerable, and the ability of various countries to deal with sea level rise will differ greatly. However projections of human mitigation strate-

Impacts of Global/Relative Sea Level Rise on Selected Coasts and Islands Worldwide

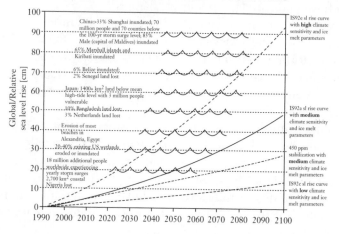

Stuart R. Gaffin, *High Water Blues*, 1997.

gies are obviously uncertain and in some cases protection measures may not lower the land-loss estimates but merely be measures to protect economically valuable structures.

Finally, even though the impacts cited are generally confined to "land-loss" estimates, as illustrated with the discussion of the small islands, a much more complex array of threats will be involved, including infrastructure losses, storm surges, salt water intrusion into both the freshwater supply and agriculture, altered tidal and wave action.

Response Strategies

Response strategies to sea level rise can be grouped into three categories: retreat, accommodate and protect.

As described in the IPCC reports, retreat strategies' measures emphasize the abandonment of land and structures and the resettlement of inhabitants. This policy would further entail preventing development in coastal areas and withdrawing of government subsidies for coastal protection. In Maine and South Carolina, for example, state legislation exists to explicitly limit how land can be developed that is vulnerable to sea level rise.

Accommodation stresses the conservation of ecosystems

with continued occupancy and adaptive management. This strategy would employ advanced planning, modification of land-use and building codes, protection of ecosystems, and hazard insurance.

Protection relies on the defense of vulnerable areas, populations, and economic activities. Hard structural options include dikes, levees, seawalls, breakwaters, floodgates, and saltwater intrusion barriers. Soft options include periodic beach nourishment, dune restoration, wetland creation, drift replenishment and afforestation. In the Netherlands, for example, dikes are built with extra elevation in order to allow for sea level rise. Similarly in San Francisco reclaimed land has to be of a certain elevation in order to allow for sea level rise.

Not all of these options are feasible for all regions. For example, heavily populated areas, island nations like Japan and seaside tourist centers probably have little choice but to protect. In many countries, the scarcity of technology and a dearth of personnel will limit the choice of accommodation options. Retreat and resettlement options will involve questions of international refugees and related disputes as well as issues of cultural traditions: To what extent will various communities be willing to resettle? How will changed or lost access to traditional fishing and hunting sites be tolerated?

Most of all though, economic considerations (resources and costs) will probably determine the feasibility of various options. In many countries, just the maintenance of existing shoreline could require substantial funding compared with the nation's GNP.

The literature on costs usually divides them into three categories: (1) capital costs of protective measures, (2) annual costs of forgone land services and (3) the costs associated with increased flood and storm frequencies. . . .

In the United States further thinking about beach protection policy is needed given the prospect of accelerated sea level rise. In 1996 the US administration attempted to limit beach nourishment projects because of the extensive US coastline (88,000 miles) and the costs now and in the projected future. Clearly US governmental policy will have to more carefully consider the need for beach nourishment programs because the impacts will only be worse in the future.

Prevention Is Necessary

Of all the forces of nature, the oceans may inspire the deepest respect and awe in many people. To see this powerful force humbled by human activity through global warming, such that its basic characteristics of sea level, coastal configuration, and (possibly) wave and storm activity are fundamentally altered must also strike a deep chord in many people. The comparative "irreversibility" of sea level rise—it will continue for many centuries even if global warming were stopped—is far longer than most other impacts from climatic change. Through greenhouse gas emissions we risk jeopardizing our complex socio-economic relationship with the sea. The rational response must be to prevent, to as great an extent as possible, dangerously high levels of greenhouse gas accumulation in the atmosphere.

"There is no *credible evidence, contemporary or historical, that a global warming will cause . . . flooding."*

Global Warming Will Not Cause Sea Levels to Rise

Richard D. Terry

According to Richard D. Terry, a marine geologist and former consultant to the U.S. Defense Department, environmental alarmists are wrong in their assertion that global warming will cause the flooding of low-lying areas through rising sea levels. In Terry's opinion, predicting changes in sea level involves the consideration of complex changes in tides, earth movements, and other oceanic and climatic processes, none of which can be accurately modeled by the computer simulations of global warming advocates. Terry claims that rising temperatures will increase evaporation and lead to more ice becoming "locked up" on the Antarctic Ice Sheet, a process that may lower sea levels by about one foot. Scientists are scaring people with the threat of flooding in order to promote a political agenda of population reduction, in the author's opinion.

As you read, consider the following questions:
1. What is the biggest problem with the accuracy of tide-gauges, in Terry's opinion?
2. According to the author, what do global warming advocates generally assume is the largest potential source of water to raise sea levels?
3. Is the Arctic melting or cooling, in Terry's opinion?

Reprinted, with permission, from "There's No Truth to the Rising Sea Level Scare," by Richard D. Terry, *21st Century*, Summer 1998, pp. 66–72. Website: www.21stcenturysciencetech.com.

Global warming proponents are sounding the alarm about potential flooding of low-lying coastal areas and low islands, but the likelihood of any global warming causing a catastrophic rise of sea level is nonexistent. As I shall show, there is *no* credible evidence, contemporary or historical, that a global warming will cause such flooding. Further, claims by global-warming modellers that they can predict sea level are a myth.

Many Processes Affect Sea Level

The processes that control or affect sea level and the origin and nature of sea level changes are complex. The ocean's surface is in constant motion and undulates. Water moves in some general direction, but the flow is turbulent and superimposed on the general movement. It is generally thought that there is a correlation between glacial lowering of sea levels and the ice tied up on the Earth's surface. Variations of atmospheric CO_2 levels and ocean temperatures are related to changes in ice volumes, and probably contribute to glacial-interglacial cycles. But, as I show, warming at the poles means more ice, not a rise in sea level. (And note that, contrary to the propaganda, we are now at the end of an interglacial.)

Tides, which are one of the indicators of ocean levels, are difficult phenomena to measure and compare. In some parts of the world there are no tides or tidal currents; in other places, tides exceed 50 ft. Tide-gauges record sea levels throughout the world, although records are limited prior to 1900. These tide-gauges are not well distributed around the world, and the records are usually irregular, requiring statistical analyses to compare any two stations. The biggest problem is that most tide-gauges are on unstable foundations; no known place on Earth is free from Earth movement. Therefore, no completely satisfactory data exist to measure or compare relative sea levels.

Solid-Earth processes that affect sea level come in many varieties: Earth movements, geological faulting, vertical movements caused by earthquakes, sea-floor uplift and subsidence, sea-floor topography, volcanism and thermal effects (super plumes, sea-floor emanations, Earth degassing), changes of land and ocean areas, sedimentation and compaction, isostasy, geoidal effects, Earth pulsations and cycles, and astronomical forces. Movements of the Earth's surface can be exceedingly large.

Other processes are oceanic and climatic: glacial surges and ice melting; climate effects (drastic weather changes that occur randomly); ocean effects (steric ocean response, temperature and salinity), long-period tides, shelf-waves and seiches, gravity waves, and others. Most of these processes are poorly understood and difficult to model, because they are not linear.

Dubious Assumptions

Global warming "predictions" are actually based on dubious assumptions, unsupported by measurement or testing. For example, global-warming advocates *assume* that they can accurately model climate and forecast sea level. But, can they? . . .

Climate modellers *assume* that the atmosphere behaves in a linear, non-turbulent, fashion. They must do so, because otherwise they cannot possibly model in detail the atmosphere or the oceans, both of which are chaotic and nonlinear. Nonlinear forces operate throughout the universe and have long haunted physicists, oceanographers, and astronomers. . . .

The major problem in simulations is that they hardly ever mimic the "real world," which is bewilderingly complex and still has many unknowns. For example, models have difficulties with: the effects of rainfall on vegetation and soils, the growth and shrinkage of sea ice, combining climate and ocean circulation, and variations of energy from the Sun, especially cloud cover. One climate model shows Death Valley filled with water! In another, oceans are modelled as a "swamp.". . .

Sea-Level Predictions Elude Modellers

Now, on to predicting sea level.

Global warming modellers *assume* that they can predict sea level—and that they can do so with breathtaking precision. Of course, this implies that modellers are able to take into account *all* the aspects of the Earth and ocean processes noted above. Earth scientists agree that predicting ocean volume changes and sea levels are difficult and, as will be discussed later, sea levels are barely measurable, and the predicted changes are well within sea-level "noise" range. In the final analysis, when it comes to the Earth sciences, including oceanography and geophysics, global warming modellers are out of their milieu.

Nonetheless, global-warming proponents *assume* that the United Nations climate models are accurate, thus permitting them to make accurate sea-level predictions. The difficulty in assessing sea-level modelling values is that the modellers present us with a moving target; that is, their sea level predictions keep changing.

One study states that there will be a rise in sea level of 10 feet by the year 2024. Elsewhere, we are told that a 4°C rise in temperature would cause sea level to rise 2 meters (m) in 500 years. In 1980, global-warming prognosticators estimated a 25-ft rise of sea level over the next 150 years. The 1985 Intergovernmental Panel on Climate Change (IPCC) report projected a "best estimate" rise of sea level of 3 ft. In the same year, a report by the U.S. National Research Council, chaired by M.F. Meier, also reduced the projected sea level rise to 3 ft. Then, in 1989, Meier, allowing more water vapor and other factors, calculated that sea level in 2050 would rise about 1 ft.

As for the IPCC, in 1989, its estimate of rise of sea level dropped to 1 foot. Then, in 1990, the IPCC report projected a "best estimate" of 66 centimeters (cm) for sea level rise in the twenty-first century.

Continuing Confusion

By 1992, however, other scientists were predicting that sea levels would *fall* by –1 ft., also as a result of global warming. A Canadian-American team of scientists reported that ice sheets will grow in size as a result of more water being tied up as snow, causing sea level to drop in the twenty-first century. At the same time, others predicted that on the basis of a forecast of a 6° to 8°F rise in temperature, sea level would rise 1 to 3 ft, as a result of the thermal expansion of the oceans. . . .

- Great caution (if not skepticism) should be given to any predicted sea level.
- Tide-gauges records are extremely variable, owing to

Earth movements. . . .

- Claims by global-warming modellers that they can pre-dict sea level are not real. In a word, predicting sea level is well nigh impossible.
- There is *no* credible evidence that global warming will cause flooding of low-lying areas. Ten years ago, when global warming alarms first sounded, had policy-makers built sea walls at great expense to protect coastal areas, it would have been a totally wasted effort. . . .

Where Does the Water Come From?

Now, we must understand that 97 percent of the water on Earth is in the ocean. If one wishes to raise the ocean's level (sea level, that is), a tremendous amount of water must be found and put into the ocean. It is generally *assumed* that the largest potential source of water to raise sea level is glacial ice. Most climate models today, however, foresee increased precipitation. If that were to happen, as we shall see, there is a good chance that sea level will *drop* as much as 2 ft in the twenty-first century.

' Why? Because increased evaporation locks up more water and puts more ice on the Antarctic ice sheet.

The Arctic Ocean has a deep ocean basin that is covered entirely by floating sea ice (frozen sea water). The density of sea ice is 0.92 grams per cubic centimeter. The temperature at which sea ice freezes is –1.9°C; salt lowers the freezing point of water. The colder the solid form gets, the less dense it becomes. Sea ice floats because it is less dense than when it is liquid form and, once frozen, ice occupies 10 percent more space. This means that melting of sea ice does *not* cause sea level to rise; it actually lowers (local) sea level.

A Melting Arctic?

It has been claimed that ice in the Arctic is melting; however, after analyzing 27,000 temperature readings, Professor Jonathan Kahl found a statistically significant trend in the opposite direction—today the Arctic is cooling. Both the Greenland and Antarctica ice caps have been growing in recent years.

More than 90 percent of all ice is stored on the Antarctic

continent; Greenland accounts for only 5 percent, and glaciers the remainder. The quantity of water stored in glaciers is debatable, but certainly is insignificant in any asserted impact on the oceans.

Sea Levels over the Past 150,000 Years

This approximate curve of sea levels is based on various measuring methods. Similar curves have been made from carbon-14 dates and the depths at which coastal fossils were found on the continental shelves.

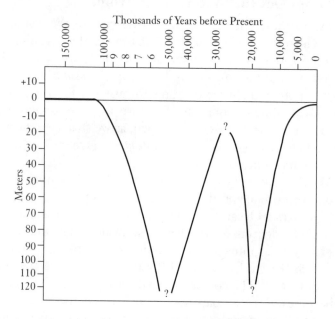

Adapted from F.P. Shepard and J.R. Curray, 1967. "Carbon-14 Determination of Sea Level Changes in Stable Areas." In *Progress in Oceanography, Vol. 4* (Oxford: Pergamon), p. 283.

Global warming will have *no* effect on the Antarctica ice cap. There are several reasons. Melting of ice on any continental ice sheet takes place only at the bottom, where it is warmed by geothermal heat. The top of an ice sheet is cold −50°C) and dry. Even with substantial heating, ice would not melt, because of its large thermal response time. The ice cap is thick, and ice itself acts as an excellent insulator, protecting it from melting.

Even if the air temperature rose, say 6° to 7°C, the ice cap would still have a temperature of ~ –46°C, and the Ice Sheet would remain solid. The air temperature above the ice sheet must reach 1°C before ice would begin to melt, and it would take +5,000 years to melt the ice cap—*if* global warming could cause the temperature to rise this much.

It is generally accepted that the rise of sea level in post-glacial time required melting of ice at a rate of 5,000 km^3 for 7,000 yrs.

More Ice from Rising Temperatures

As the air temperature heats up, it holds more water vapor. This is the opposite of the theory of global warming's basic assumptions. If the mean air temperature around Antarctica were to rise, more clouds would develop; more clouds would cause the air temperature to fall. Water evaporated from the oceans would accumulate as snow and become "locked up" on the ice sheet. Therefore, the ice sheet would thicken. This process would lower sea level by about 1 foot.

From all this, we can confidently say that global warming's basic argument—*warming will cause sea level to rise*—is completely at odds with the dynamics of the hydrologic cycle. This self-regulating process, which restores equilibrium, is a well-known principle that every freshman college chemistry student learns.

It often has been stated that, if melting of the Antarctic ice cap took place, sea level would rise ~150 ft, a figure widely reported by the media. This is a lie, as already mentioned. The Antarctic Ice Sheet will *grow*, rather than diminish, if temperatures increase in the twenty-first century, accumulating snow faster than it loses ice. Antarctica has little meltwater, owing to the extreme cold, but a small amount reportedly reaches the ocean from the East Antarctic Ice Sheet.

If the Antarctic Ice Sheet were to completely melt, the air temperature of Antarctica would have to be 1°C or higher, over thousands of years. Not only that, but in order to get the temperature of Antarctica to rise to 1°C, the entire atmosphere would have to have a temperature increase of ~51°C–210°F. (Imagine Washington, D.C., summer temperatures of 210°F. . . .)

For these reasons, the Antarctic Ice Sheet can obviously be ignored in global warming scenarios. No one expects the melting of the Antarctic Ice Sheet, even with a 7-fold increase in CO_2.

Ruling Out Glaciers and Thermal Expansion

This leaves the global warming flood propagandists with only Greenland, mountain glaciers, and icebergs, all of which are trivial sources of water. As Tom Wigley summed up the problem of modelling sea level rises from glacier melting, "Wide uncertainties still remain." And, as glaciologist F.B. Wood has pointed out, "if there were a magic way to melt all the land glaciers of the Northern Hemisphere, sea level would rise only 10 cm."

Ah, but if we can't raise sea level by melting the ice, global-warming advocates then pull out of their hat the fallacy of *thermal expansion* of the ocean. In theory, this could raise sea level 1 to 2 ft, but, as we have seen, such a small rise would not be apparent. And then, too, it would take about 13,000 years for the action to take place.

Why Global Warming?

Given the absurdity of the claims of global warming propagandists about ice melt, why do they persist in scaring people about rising ocean levels? My conclusion is that it's purely political, and has to do with population reduction.

Global-warming gurus have built careers and fortunes warning people that sea level is rising. These fear-mongers feed on the public's lack of knowledge about the true facts. They counsel people living in low-lying coastal areas—usually with the help of a pliant and ignorant media—that they are in danger of being inundated by a rising sea. These gurus have argued that a rising sea level is already demonstrated by the wide oscillations of lake levels in the Caspian Sea. (A Russian geologist, however, has shown that the fluctuations are caused by tension and compression in the Earth's crust.)

People living in the coastal zone are being frightened into thinking that they are about to lose everything. They are told that they can expect higher-than-normal tides and storm surges, El Niño events, hurricanes, tidal waves, and the like. The media—TV, newspapers, even pseudoscientific publications—use archival films and photographs showing calamitous ocean and climatic events, passing them off as if they happened a few days ago.

Australian scientist Peter Sawyer characterized the situation this way: "It's a bit hard to reduce people to a state of fear and panic with the 'threat' of more food and better climatic conditions [from warmer temperatures], so something else had to be found. It's a measure of just how flimsy the whole greenhouse argument is, that the worst 'threat' that could be presented was that oceans-levels will somehow rise, and flood out some coastal areas."

It's time for people to wake up, realize the serious consequences stemming from the policies of global warming hacks and bureaucrats, and fight back with the truth. The real global warming catastrophe is how easy it is for some scientists to scare people with scenarios that have no scientific validity.

Periodical Bibliography

The following articles have been selected to supplement the diverse views presented in this chapter. Addresses are provided for periodicals not indexed in the *Readers' Guide to Periodical Literature*, the *Alternative Press Index*, the *Social Sciences Index*, or the *Index to Legal Periodicals and Books*.

Seth Borenstein — "Experts See Global Warming Disaster," *San Diego Union-Tribune*, February 19, 2001.

David Bramley — "Weather, Climate, and Health," *World Health*, September/October 1998.

Grover Foley — "The Threat of Rising Seas," *Ecologist*, March/April 1999.

Ross Gelbspan — "The Global Warming Crisis," *Yes!*, Winter 1999/2000.

James Glanz — "Droughts Might Speed Climate Changes," *New York Times*, January 11, 2001.

Bob Herbert — "Adjusting to the Earth's Weather Extremes," *New York Times*, September 24, 1999.

Mark Lynas — "Storm Warning," *Geographical*, July 2000.

Lewis MacAdams — "In a Summer of Fire, Is Warming a Cause?" *Los Angeles Times*, August 27, 2000.

Thomas Gale Moore — "Happiness Is a Warm Planet," *Wall Street Journal*, October 7, 1997.

David Nicholson-Lord — "The Drowning of the Earth," *New Statesman*, March 6, 2000.

Kelly Reed — "To the Extreme," *Greenpeace*, Winter 2000. Available from 702 H St. NW, Washington, DC 20001.

S. Fred Singer — "The Sky Isn't Falling, and the Ocean Isn't Rising," *Wall Street Journal*, November 10, 1997.

Tomari'i Tutangata — "Rising Waters, Falling Hopes," *Toward Freedom*, November 2000.

John Noble Wilford — "Ages-Old Icecap at North Pole Is Now Liquid, Scientists Find," *New York Times*, August 19, 2000.

John Noble Wilford — "Open Water at Pole Is Not Surprising, Experts Say," *New York Times*, August 29, 2000.

CHAPTER 4

Should Measures Be Taken to Combat Global Warming?

Chapter Preface

In December 1997, world leaders met in Kyoto, Japan, to continue negotiations on a United Nations treaty drafted in 1992 to address global warming through the reduction of greenhouse gas emissions by industrialized nations. The negotiations resulted in the Kyoto Protocol, which requires the United States to reduce its greenhouse gas emissions by 7 percent below 1990 levels between 2008 and 2012. Thirty-seven other industrialized nations also agreed to reductions. Although a Clinton administration official signed the Kyoto Protocol on November 12, 1998, the U.S. Senate must ratify it to render it a binding agreement.

A debate has arisen over whether ratification of the Protocol will cause more harm to the U.S. economy than would occur if global warming were left unchecked. Those in favor of the Protocol believe that it will reduce the threat of climate disasters associated with global warming and that the economy will benefit from the creation of alternative energy industries. Explains the environmental journalist Ross Gelbspan, "A global transition to climate-friendly energy sources would substantially expand the . . . total wealth in the global economy . . . through the large-scale creation of jobs."

Those opposed to the Protocol, however, contend it will lower American living standards. Anti-Kyoto forecasters predict rising gas, food, and housing costs as a result of a reduced energy supply. In addition, others complain that the United States will shoulder an unfair burden for reductions, while developing nations will be exempt from emissions restrictions. According to the Competitive Enterprise Institute, "This treaty will not include legally binding emissions standards for more than 130 developing countries, including China [and] Mexico. . . . This makes no sense."

Ratification of the Kyoto Protocol is one response to the potential climate impacts of global warming. Others responses include reducing automobile usage, planting trees, and the revival of nuclear power. The authors in the following chapter debate the courses of action that will prove most effective in offsetting the effects of global warming.

"By pursuing sound domestic policies, the
United States can reach its Kyoto target at
a relatively modest cost."

The United States Should Support the Global Climate Treaty

White House Council of Economic Advisers

In December 1997, over 150 countries negotiated the Kyoto Protocol, a United Nations treaty intended to combat global warming. If ratified, the treaty would require the United States to reduce its carbon dioxide emissions by 7 percent below 1990 levels by 2012. In the following viewpoint, former president Bill Clinton's Council of Economic Advisers argues that the United States should support the treaty because it is a necessary response to the risks posed by global warming and can be implemented at a modest cost. Although President George W. Bush withdrew U.S. support for the treaty after taking office in January 2001, environmentalists continue to advocate its ratification.

As you read, consider the following questions:
1. What are the three basic kinds of flexibility provisions included in the Kyoto Protocol, according to the authors?
2. In the authors' opinion, why is analyzing the costs and benefits of mitigating global warming a difficult undertaking?
3. How much would a doubling of pre-industrial concentrations of greenhouse gases cost the U.S. economy, as reported by the authors?

Reprinted from the Executive Summary of *The Kyoto Protocol and the President's Policies to Address Climate Change*, published by the White House Council of Economic Advisers, July 1998.

The primary purpose of this viewpoint is to examine costs and benefits of taking action to mitigate the threat of global warming. In particular, we examine costs and benefits of complying with the emissions reduction target for the United States set forth in the Kyoto Protocol on Climate Change, negotiated in December 1997. For reasons discussed in this viewpoint, it is our conclusion that, with the flexibility mechanisms included in the treaty, and by pursuing sound domestic policies, the United States can reach its Kyoto target at a relatively modest cost. Moreover, the benefits of mitigating climate change are likely to be substantial.

Before considering the economics of taking action, however, we ought to step back and ask the threshold question—whether taking action to mitigate global climate change is necessary in the first place.

The Rationale for Taking Action

The great weight of scientific authority suggests that climate change is a serious problem and that prudent steps to mitigate it are in order. In essence, we need to take out an insurance policy with reasonably priced premiums. As long ago as 1992, the National Academy of Sciences, in a study entitled *Policy Implications of Greenhouse Warming*, concluded that ". . . even given the considerable uncertainties in our knowledge of the relevant phenomena, greenhouse warming poses a potential threat sufficient to merit prompt responses. . . . Investment in mitigation measures acts as insurance protection against the great uncertainties and the possibility of dramatic surprises."

What the science tells us is that greenhouse gases are rapidly building up in the atmosphere as a result of the burning of fossil fuels and deforestation; that the concentration of these gases is 30 percent higher than it was at the beginning of the industrial revolution; and that this concentration is expected to reach almost twice current levels by 2100—a level not seen in 50 million years. Theory and computer models suggest that this increased concentration of greenhouse gases could warm the Earth by about 1.8 to 6.3° Fahrenheit (F) by 2100. By way of comparison, the last ice age was only about 9° F colder than today. Moreover, much evidence suggests that warming is already underway. For example, we know from ice cores and other data that we are living in the hottest century since at least 1400, that the nine hottest years since records were first kept in the late 19th century have all occurred since 1987, and that 1998 is the hottest year on record.

Scientists predict a range of likely effects from global warming. For example, the rate of evaporation is expected to increase as the climate warms, leading to increasingly frequent and intense floods and droughts. Sea level is projected to rise 6–37 inches by 2100. A 20-inch rise could inundate about 7,000 square miles of U.S. territory. Warmer temperatures would be expected to increase the risk of mortality from heat stress, aggravate respiratory disease, and increase the range and rates of transmission of some infectious diseases.

Scientific opinion is not unanimous on these points, but most independent climate scientists believe that global climate change poses real risks. A few scientists contest the notion that increasing concentrations of greenhouse gases will warm the planet, while a few others concede that the earth is indeed getting warmer, but argue that this is a good thing—"a wonderful . . . gift from the industrial revolution," in the words of one. But these are distinctly minority views. The prevailing view is that the risks of climate change warrant prudent and prompt action. Prompt because to wait for greater scientific certainty could have very large costs. Greenhouse gases are long-lived and the decisions being made by governments and firms in the next decade with respect, for example, to the kinds of power plants to build or the kinds of energy sources to develop, are likely to have significant consequences for our ability to limit the buildup of greenhouse gases.

Consequently, there is a substantial rationale for acting now. Our task is to act in a manner that responds appropriately to the scope of the risk while at the same time being economically sensible.

In October 1997, [then] President Clinton announced a domestic program designed to reduce greenhouse gas emissions. In essence, the program contemplated (a) a set of activities that made sense as good energy and environmental policy irrespective of whether an agreement were reached in Kyoto, and (b) a mandatory domestic emissions trading system that would take effect in the 2008–2012 period if an agreement in Kyoto were reached and approved by the U.S. Senate.

The Kyoto Protocol

The Kyoto Protocol, which requires the advice and consent of the Senate, commits industrialized nations to take on binding targets for greenhouse gas emissions, and includes three basic kinds of flexibility provisions that were proposed by the United States. These provisions—commonly referred to as "when", "what", and "where" flexibility—have great potential to significantly lower the costs of meeting the Kyoto targets. "When" flexibility appears in the form of a multi-year commitment period (2008–2012), and allowance for "banking" of emissions reductions. The freedom for countries or companies to delay or accelerate reductions within an agreed upon time frame can help lower costs. "What" flexibility is provided by both the inclusion of all six greenhouse gases—enabling reductions in emissions of one gas to be used to substitute for increases in emissions of another—and the coverage of certain "sink" activities, such as afforestation or reforestation, that absorb carbon. Most important, the Protocol incorporates "where" flexibility in the form of international emissions trading and joint implementation among countries that take on binding targets, coupled with a "clean development mechanism" allowing industrial countries or firms to earn credits for projects in the developing world that reduce emissions. These mechanisms can provide opportunities for industrial countries and firms to secure low-cost reductions and for developing countries to achieve sustainable growth.

Developing countries did not take on binding emissions targets at Kyoto, although they did agree to provisions for the Clean Development Mechanism. President Clinton said that he would not submit the Protocol to the Senate without meaningful participation from key developing countries. While the Clean Development Mechanism provides a down payment on such participation, the Clinton

Administration actively sought greater developing country engagement.

Analyzing the costs and benefits of mitigating climate change is a difficult undertaking for three reasons. First, uncertainties remain about significant details of certain provisions in the Protocol. Second, available models have inherent limitations in their abilities to analyze even short-term costs and benefits. Third, it is extremely difficult to quantify the long-term economic benefits of climate change mitigation. Thus, while we have summarized the literature, we have not calculated a monetary value of these benefits.

Local Efforts to Reduce Global Warming

In 1999, the U.S. Senate voted 98-0 to reject any climate change treaty that does not require poor nations to accept similar pollution reductions. Today attitudes in the Bush White House mirror those in the Congress, where the reactions range from deep skepticism to outright hostility.

Which means that Americans who want to meet their global responsibility will have to do it themselves—at least in the beginning. This doesn't mean simply changing individual behaviors, but rather, changing the rules to channel entrepreneurial energy and scientific genius in a direction that meets the needs of future generations.

Many cities and some states and counties have enacted resolutions encouraging greenhouse gas reductions. Most are directed at higher levels of government. None have yet translated rhetoric into significant actions.

Here's my suggestion for a powerful first step. Every city council, county commission, school board, state legislature and other tax exempt bond-issuing agencies in the country should require that any structure or piece of equipment that is financed with public money must satisfy our global obligation to reduce global warming.

David Morris, *New Rules Journal*, Spring 2001.

Recognizing these difficulties, our conclusion is that the costs for the United States to meet its Kyoto emissions target are likely to be modest if those reductions are undertaken in an efficient manner employing the flexibility measures of emissions trading (both domestic and international), joint implementation, and the Clean Development Mechanism. This would be so even without considering the direct benefits of mitigating climate change or the impact that key additional factors—such as the President's domestic climate change proposals, the ancillary benefits of improved air quality, or the inclusion of sinks—could have on lowering the net costs of mitigation.

Our conclusion concerning the costs of complying with the Kyoto Protocol is not entirely dependent upon, but is fully consistent with, formal model results. For example, given the flexibility measures noted above, with key developing countries participating in trading, and *excluding* both the benefits of mitigating climate change and the key additional factors just noted, estimates derived using Battelle's Second Generation Model (SGM) suggest that the resource costs of attaining the Kyoto targets for emission reductions might amount to $7–12 billion per year in 2008 to 2012, or just 0.1 percent of projected gross domestic product (GDP). The same model predicts that emission permits in 2010 would cost between $14 and $23 per ton of carbon equivalent—which would translate into an increase of about 4 to 6¢ per gallon of gasoline. The increase in energy prices would raise the average household's energy bill in 2010 by between $70 and $110 per year—a relatively small amount compared to typical energy price changes. Moreover, this increase would be substantially offset by the decline in electricity prices resulting from the Clinton Administration's electricity restructuring proposal.

These numbers are instructive. They demonstrate the importance of flexibility measures like emissions trading and the potential for meeting our Kyoto target at a relatively modest cost. However, it is just as important to understand what these numbers do *not* say. They do not tell us about either (a) the economic *benefits* of mitigating climate change or (b) the potential for any other domestic policy measures (aside from emissions trading) to reduce costs further and/or to increase the percentage of greenhouse gas reductions we can accomplish at home. The reason is that the SGM model we used to generate these numbers does *not*, by its terms, account for either of these factors.

There are substantial long-term benefits of mitigating global climate change. Monetary estimates of damages from the environmental, health, and economic impacts of global warming during the next century range in the tens of billions of dollars per year. One noted economist, William Cline, has estimated that a doubling of pre-industrial concentrations of greenhouse gases would cost the U.S. economy about 1.1% of GDP annually—some $89 billion a year in today's terms. Moreover, these estimates do not reflect the potential costs of so-called "non-linearities"—the risk that global warming will lead not to gradual and predictable problems, but to relatively abrupt, unforeseen, and potentially catastrophic consequences. Although we do not think the benefits of mitigating climate change are, at this stage, quantifiable with adequate precision, they are nonetheless likely to be real and large in the long run.

There are also ancillary benefits of reducing greenhouse gas emissions—in particular, the corresponding reductions in conventional air pollutants like sulfur dioxide or nitrogen oxides. These benefits alone could produce savings equal to about a quarter of the costs of meeting our Kyoto target.

The Impact of Domestic Initiatives

Following President Clinton's October 1997 policy announcement, the Clinton Administration pursued a number of domestic initiatives that will help reduce greenhouse gas emissions. These initiatives—all of which are consistent with our commitments under the 1992 Framework Convention on Climate Change, which the Senate approved that same year—could reduce costs and/or increase the amount of reductions accomplished through domestic action. First, the Clinton Administration's $6.3 billion budget proposal to promote energy efficiency and renewable energy should help increase the rate of technology development and diffusion. Many of the components of this initiative reflect recommendations made in an October 1997 report by President Clinton's Committee of Advisors on Science and Technology (PCAST), which concluded that "the inadequacy of current energy R&D is especially acute in relation to the challenge of responding prudently and cost-effectively to the risk of global climatic change. . . ."

Second, the Clinton Administration's electricity restructuring proposal was estimated to reduce greenhouse gas emissions in the United States by about 25 to 40 million metric tons per year. Competition would provide a direct profit incentive for generators to produce more electricity with less fuel and improve energy efficiency. Several specific provisions in the Clinton Administration's proposal would yield further emissions reductions.

Third, the Clinton Administration conducted industry consultations aimed at promoting voluntary agreements with major energy-intensive industries, energy providers, and others to yield further emissions reductions. One such agreement, the Partnership for Advancing Technology in Housing (PATH), announced in May 1998, established goals for voluntary improvements in home energy use that would reduce emissions in 2010 by about 24 million metric tons of greenhouse gas emissions.

Fourth, the Clinton Administration pursued an active program to reduce emissions produced by the federal government, the nation's largest consumer of energy.

As noted above, models like SGM, while well equipped to assess policies such as a tradable permit program, do not assess policies like these. To the extent that policies like these boost the rate at which energy efficiency improves, the United States could lower the cost of mitigation and increase the amount of reductions made domestically.

Finally, our illustrative analysis, based on the SGM model, did not account for the effects of carbon sinks in reducing net greenhouse gas emissions. Opportunities to reduce net emissions through carbon sinks could further reduce the costs of achieving the Kyoto target and increase domestic reductions.

The current state of the science provides a powerful rationale to take prompt, prudent action to mitigate climate change. The agreement negotiated in Kyoto includes flexibility mechanisms that will allow the United States to meet its Kyoto target at a modest cost. Additional factors not included in the modeling effort—such as President Clinton's domestic climate change policies, the inclusion of sinks, and the ancillary benefit of improving air quality—could lower costs even further and increase the percentage of reductions made through domestic action. The benefits of mitigating long-term impacts of global climate change, while not precise enough to quantify at this stage, are likely to be very important. In short, this is an insurance policy we should buy and it is one we can buy for reasonably priced premiums.

> *"Implementation of the [Kyoto protocols] . . . would have meant a $397 billion lower gross national product than if the U.S. opted out."*

Abandoning the Global Climate Treaty Is Beneficial to the United States

George Melloan

George Melloan contends in the following viewpoint that the United States, under the leadership of President George W. Bush, was correct to abandon its support for the Kyoto Protocol, a United Nations global climate treaty negotiated in December 1997 by more than 150 countries. Under the terms of the treaty, the United States would be expected to reduce its carbon dioxide emissions by 7 percent below 1990 levels by 2012. In Melloan's opinion, there is no plausible evidence of a link between greenhouse gas emissions and global warming, and U.S. compliance with the treaty would weaken the economy due to the exorbitant costs associated with replacing carbon-intensive fuels. Melloan is a columnist at the *Wall Street Journal.*

As you read, consider the following questions:
1. What happened to a 1996 report drafted by UN scientists, in Melloan's opinion?
2. How would U.S. compliance with the Kyoto treaty have affected the country's gross national product, in the author's opinion?

Something akin to mass hysteria has erupted in certain circles over George Bush's announcement that the U.S. will not support the Kyoto protocols designed to defend us all against "global warming." His reason was that the United Nations plan could put a sputtering U.S. economy into the tank. He could have added a second reason: There is no plausible evidence that a significant global-warming trend exists.

"Gobbledygook" Replaces Science

The scientific argument on this issue ended, for all practical purposes, somewhere in the mid-1990s. The scientists on the U.N.'s own Intergovernmental Panel on Climate Change (IPCC) were so skeptical in a 1996 draft report that their political betters chose to censor them. The pols substituted gobbledygook for the scientists' admission that they could find no clear evidence of a link between temperature change and greenhouse gases. Dr. Frederick Seitz, former head of the U.S. National Academy of Sciences, fumed that he had never seen "a more disturbing corruption of the peer review process . . ."

By the year 2000, "global warming" had become a joke in America, finding its way into cartoons and the repertoires of late-night jokesters. Al Gore, one of the original "global warming" Chicken Littles, didn't choose to stress his role in producing the Kyoto protocols during his presidential campaign. Obviously, enviroscares were losing some of their sex appeal as more and more Americans began to wonder what was so bad about warmer weather. They would wonder even more when one of the coldest winters on record descended on the Northeast, piling up five-foot snow banks.

Politics and Public Fears

But the Kyoto juggernaut just kept on rolling in other parts of the world. German Chancellor Gerhard Schroeder, a White House visitor in March 2001, professed shock at the Bush announcement. He, after all, is in a coalition with the Greens, who already are losing favor with German voters. French President Jacques Chirac is trying to fend off a scandal investigation in Paris and no doubt hoped to get some

Green support by calling the Bush statement "disturbing and unacceptable."

What's causing all of this deep concern among politicians is not the fear that the earth will be incinerated next July, but the prospect that the "environmental movement" may finally be fizzling out as a political constituency. It has made itself a powerful political force by stirring up public fears of mostly imaginary dangers, such as global warming or genetically modified organisms (GMOs). The U.N. got by in the 1980s with the costly and scientifically dubious Montreal Protocol, which outlawed Freon as a threat to the ozone layer. Enviros pretty much killed off nuclear energy with regulatory delays that forced up the cost of new construction. But Kyoto may have been a bridge too far.

Taxing Mother Nature

One of the many problems with Kyoto is that it targeted Mother Nature herself. In some theoretical models, it was alleged that increasingly abundant "greenhouse" gases, principally carbon dioxide and methane, would trap heat and raise the temperature on earth. It was thus deduced that industrial emissions of "greenhouse gases" must be regulated and somehow curtailed.

The first problem with this theory was that there was no empirical evidence to support it. Measuring the temperature of the earth is no easy matter, but measurements showed that most of the barely noticeable temperature rise of the past 100 years occurred before 1940, when the globe was far less industrialized than today. Serious scientists also pointed out that the greenhouse gases are a fundamental part of the biosphere, necessary to all life, and that industrial activity generates less than 5% of them, if that.

Despite all the scientific doubts, the U.N. political bandwagon rolled on toward the goal of taxing carbon-dioxide emissions. Developing countries got an exemption because of their quite plausible claim that a carbon tax would retard their development. A system of tradable permits was proposed so that countries that have suffered industrial decline, like Russia, could sell emission rights to industries with emissions over the target limits. No one has figured out how

Estimates of Increases in the Average Cost of Energy Under the Kyoto Protocol

	U.S. Department of Energy	WEFA, Inc.	White House
Carbon Permit Prices	$348 per ton in 2010. Increased energy costs for average households: $1,740 annually.	$265 per ton in 2010.	$14 to $23 per ton in 2010. Increased energy costs for average households: $70 to $110 annually.
Electricity Prices	11 cents per kilowatt-hour (kWh) in 2010. Increase: 86.4% over baseline of 5.9 cents per kWh.	9.8 cents per kWh in 2010.	6.1 cents to 6.2 cents per kWh in 2010. Increase: 3.4% to 5.1%.
Gasoline	$1.91 per gallon in 2010. Increase: 52.8% over baseline of $1.25 per gallon.	$1.83 per gallon in 2010.	$1.29 to $1.31 per gallon in 2010. Increase: 3.4 cents to 5.5 cents per gallon, or 3% to 4%.
Fuel Oil	$1.90 per gallon in 2010. Increase: 76% over baseline of $1.084 per gallon.	$1.89 per gallon in 2010.	$1.14 to $1.17 per gallon in 2010. Increase: 4.8 to 7.8 cents per gallon.
Natural Gas	$9.57 per thousand cubic feet (mcf) in 2010. Increase: 147% over baseline of $3.87 per mcf.	$7.61 per mcf in 2010.	$4.00 to $4.13 per mcf in 2010. Increase in cost 5.3% to 8.7%.

Alexander F. Annett, *Heritage Foundation Backgrounder*, October 23, 1998.

to make this ill-defined scheme work, but there already is a budding market in "emission rights" in Europe, as alert traders sense there might be money in the U.N.'s folly.

Bush Balks at Costs

What stopped George Bush and Congress was the estimate of what the U.N. project would cost the U.S. Al Gore agreed at Kyoto in 1997 that by 2012 the U.S. would reduce its carbon emissions to 7% below 1990 levels. The only way to cut carbon-dioxide emissions is to replace carbon-intensive fuels, which rules out cheap fuels, like coal. The U.S. Department of Energy estimated that implementation of the Gore promise would have meant a $397 billion lower gross national product in 2010 than if the U.S. opted out. Kyoto would boost electricity prices by 86.4% and other energy costs accordingly.

Mr. Bush no doubt concluded that an American president would be out of his mind to commit to something like this while the economy is slowing and California is starving itself to death with price controls on electric energy. Besides, all the American enviros had voted for Al Gore or, more likely, the man who is responsible for a lot of this nonsense, Ralph Nader.

One of the delightful things about the American economy, of course, is its ability to absorb shocks through its sheer size and adaptability. Kyoto has had one salutary effect, reviving interest in nuclear power. Despite all the abuse it has taken from the Naderites, nuclear power had its best year ever in 2000. No new plants were built but owners have been recommissioning existing plants and upgrading their generating capacity with new equipment and instruments. Nukes don't create carbon. They would be an attractive source of power even if Kyoto didn't exist, because of their cleanliness and relative safety.

Farewell to a "Fool's Errand"

One reason for the shock at the Bush announcement is the perception by the enviro-scaremongers that the old magic has stopped working. Without the U.S. on board, Kyoto will become a worthless relic. What will all the people who worked to put it together do then? Invent another global threat? A lot of careers are wrapped up in this treaty and a lot of hopes for bureaucratic jobs that would be created trying to make the nightmare of tradable permits work.

Think of the power of having the whole world dancing to a U.N. tune! All lost. George Bush may well have had his finest hour when he summoned up the courage to tell other national leaders that they have been on a fool's errand.

"Our stock of motor vehicles is not only related to rising temperatures and erratic weather but a parent of the problem."

Curtailing Automobile Use Would Reduce Global Warming

Jane Holtz Kay

Jane Holtz Kay contends in the following viewpoint that carbon dioxide emissions from cars and trucks are the largest single contributor to global warming. She argues that the cost of America's dependency on the automobile is enormous in terms of environmental damage, road construction and maintenance, excessive energy consumption, and lives lost in car accidents. Raising the tax on gas to curtail car usage and building less car-dependent urban areas are proposals that Kay advocates to mitigate the automobile's negative impact on global warming. Kay is the author of *Asphalt Nation: How the Automobile Took over America and How We Can Take It Back.*

As you read, consider the following questions:
1. What type of vehicle is to blame for a 3.4 percent increase in U.S. emissions, according to the Atmosphere Alliance as cited by the author?
2. According to Kay, how many pounds of carbon dioxide are added to the atmosphere every second by America's motor vehicles?
3. What percentage of the U.S. gross domestic product is committed to highway-based transportation expenditures, in Kay's opinion?

Reprinted, with permission, from "Infernal Combustion," by Jane Holtz Kay, *In These Times*, August 8, 1999.

You don't need a weatherman to tell that the whole earth has become the scorched earth. And you don't need a climate course to teach you that the temperature has become hot news. In the hottest decade of the millennium, "severe weather alerts" are as constant as the calendar.

It started during the winter of 1998 with the headlines: "South Gets White Christmas and Loses Power" and "California Farmers Hope to Salvage Some Citrus." It continued with blizzards in the Midwest, tornados in Florida, and hot-to-warm climate quick steps in New England. By late spring, the Los Angeles cool and the East Coast steam had reversed the natural order of the continent.

But if weather scares have chilled us out and heated our consciousness, there is one thing that the fluctuating thermometer and rising tides don't record. And that's the complicity of the car. Whatever the assessment of the damage of the capricious climate, the political and financial barometers have yet to register the largest single contributor to global warming.

The Cost of Auto-Dependency

"Is your current car too closely related to the fossil fuel it burns?" asks an advertisement for a luxury automobile. You bet it is. Our stock of motor vehicles is not only related to rising temperatures and erratic weather but a parent of the problem. In just one example, the Atmosphere Alliance has blamed a sharp jump of 3.4 percent in U.S. emissions—more than the total of most nations—on one automated energy hog, the sport-utility vehicle (SUVs).

But SUVs on steroids are just the newest phase of U.S. auto-dependency. Clock the minutes: Every second the nation's 200 million motor vehicles travel 60,000 miles, use 3,000 gallons of petroleum products and add 60,000 pounds of carbon dioxide to the atmosphere—that's two-thirds of U.S. carbon dioxide emissions.

The surprise is that despite the motor vehicle's role in making the weather gyrate like a Dow Jones graph, the total cost of America's auto-dependency remains a dirty but hidden secret. The roads we build to serve the car, the fuel we extract, the industrial energy consumed in producing 15 million motor vehicles a year are enormous—and largely un-recorded. U.S. cars and trucks carry three-quarters of a tril-lion dollars in hidden costs. Often lacking a dollar sign, their tally ranges from parking facilities to police protection; from registry operations to uncompensated accidents. Cars bought on the installment plan drive up consumer debt by 40 per-cent, making the General Motors Acceptance Corporation the largest consumer finance institution in the world. And we haven't even calculated the environmental cost of global warming in repairing the damage from floods and fires.

How do we right this equation? We need to acknowledge the exactions of our auto-based existence. The love affair with the motor vehicle that festoons our policy like a GM hood ornament comes at a steep price, personally, socially and environmentally.

Beyond the $93 billion a year that local, state and national governments spend on roads, we must tally other expendi-tures, from the 41,500 lives lost annually in car accidents to the automobiles and auto parts that account for two-thirds of our trade deficit with Japan. From 8 billion hours a year stuck in traffic to the $100 billion a year spent on the mili-tary budget defending our Middle East oil supply, the visible and invisible costs of the car mount. Count, too, the rising cost of that oil extraction as we labor to clean or discover new reserves, which are predicted to dwindle and become pricier on their way to exhaustion.

Driving Sprawl

Beyond building and running our cars, there is the environ-mental and financial toll of car-bred sprawl. The land bull-dozed into asphalt is a lost opportunity cost. The wetlands and farmland paved (two million arable acres a year), the open space or city split by an arterial highway, or the hilltop sprout-ing the four-leaf clover interstate is, to say the least, a minus.

Then there are the other invisible losses. The price of car-bred infrastructure subsidized to take us to the sprawling edges demands, in turn, an evermore pricey and energy-squandering infrastructure of electricity, cable and sewage lines at the end of the road. Consider that by laying asphalt for the automobile, we give over more than half our cities to roads and parking lots. (Note how each automobile demands seven spaces to move and park—one at home, one at work, one at the mall and four for the road network.) Chart our subsidies for such incidentals as parking: for one, the 85 million employees given free spaces whose real estate is worth an average of $1,000 apiece. This amounts to an $85 billion lure—and $85 billion denied to non-drivers. No other country carries our loss in property taxes from such "investments." Finally, compute the price of 4,000 "dead" malls and countless Main Streets languishing in the wake of the highway-based exodus. The tragic loss of community cannot be reckoned.

Reducing Emissions with Fuel-Efficient Vehicles

The Kyoto treaty on global warming calls upon the United States to reduce its carbon dioxide emissions by seven percent in just over ten years' time. While this seven percent reduction may seem daunting, a host of initiatives are already under way that could get the U.S. most of the way toward compliance.

One such effort is the campaign to create more fuel-efficient automobiles. In the fall of 1997, the Department of Energy, the Los Alamos National Laboratory, and Arthur D. Little Co. announced they had developed a fuel processor capable of extracting hydrogen from gasoline. This technology could allow efficient conversion of gasoline and other fuels to electricity, opening the way for a new generation of fuel-efficient, electrically driven motor vehicles that might be available for mass markets within a decade.

Michael B. McElroy, *The New Republic*, May 4, 1998.

What false economy allows us to dismiss these debts? To simply credit highway-based transportation as 18 percent of our gross domestic product—more than health and education combined? What perverse sense of the environmental balance sheet lets us tamper with the fate of the planet without noting these debts? In the end, our pro–fossil fuel government and industry underwrite the car culture that undermines planet preservation. It favors the private car vs. public transportation at seven-to-one, offers single-family mortgages and policies that undercut core cities and suburbs, and gives the highway men a free lunch on a silver platter.

Breaking the Car Habit

Curbing the car to protect the climate is good financial as well as environmental policy. Making the car pay its way by altering pricing policies to stop the subsidies would reduce costs while cutting fossil fuel. Raising the tax on gas—or on carbon dioxide–spewing gas guzzlers or on number of miles driven—would lessen auto use and impact. So would congestion pricing, tolls and parking fees. It is time to follow the other industrialized nations of the world by raising gas prices to $4 or $5 a gallon, funnelling these funds to good public transportation and lessening the need for autos in the first place.

Changing sprawl-inducing land patterns that have made two or three cars a (perceived) prerequisite in half our households is also essential. By reinvesting in public transportation, good planning, mixed-use zoning and other improved land use policies, we create dense neighborhoods and urban infill for the clustered physical environment that supports the mass transit, trains, bicycling and walking that will ease us out of the car trap.

None of these routes to reduce auto-dependency and halt global warming is built in a day, but they can begin instantly on a personal and political level. As Washington and Wall Street slouch their way to climate protection, we need to do more—far more—than give lip service to this mindset. "Cogito Ergo Zoom" is how *Automobile* magazine describes America's attitude to the internal combustion machine. More cogitating and less zoom would be better. So would activism from the bottom and leadership from the top to replace a mentality as stuck in traffic as our way of life. The Atlantic would be rolling across the Adirondacks and the glaciers melting into Miami before the people who brought us the *Exxon Valdez* and the Corvair flipped the switch on their course to stop climatic upheaval. It is time for the rest of us to brake the automotive gluttons that fuel global disarray.

"*Zealots . . . are convinced by the openly dubious science regarding global warming and . . . have also managed to delude the public that the [automobile] industry is dominated by churlish profiteers bent on destroying the environment.*"

Automobiles Are Not Contributing to Global Warming

Car and Driver

In the following viewpoint, the editors of *Car and Driver* magazine argue that contrary to the claims of environmentalists, gasoline-burning vehicles contribute only a small percentage to total greenhouse gas emissions. Alternative energy sources such as fuel cells, electric cars, and solar power cannot compete with gasoline-powered engines on the basis of cost and fuel efficiency, according to the authors. In the authors' opinion, curtailing automobile use is a misguided approach to solving the dubious problem of global warming.

As you read, consider the following questions:
1. How has the automobile industry reacted to the anti-automobile media campaign, in the authors' opinion?
2. According the authors, what percentage of passenger vehicles sold are sports utility vehicles, pickups, and vans?
3. By how much would total carbon dioxide emissions from man-made sources drop if all gasoline-burning vehicles were eliminated, according to James Johnston as cited by the authors?

Reprinted, with permission, from "Lost in the Ozone Again," editorial in *Car and Driver*, vol. 44, no. 1, July 22, 1998.

A mong the precious few benefits to accrue from the spate of outrages committed in Washington during 1998 is the disappearance from the headlines of the Kyoto conference and its attendant anti-car pedantry. You may recall that before [former] President Bill Clinton's endless rutting and his whey-faced squirming to escape blame captured the attention of the nation, [former vice president] Ozone Albert Gore was about to launch a frenzied attempt to ride a bubble of global warming into the White House in the year 2000. A centerpiece of that campaign was the demolishment of the hated automobile culture.

Poor Albert seems dogged by bad luck. Not only was he tainted by the campaign-funding scandals while his smarmy boss danced free, but his grand plan built on environmental fear mongering and anti-car hysteria (based, apparently, on the simple-minded premise—supported by the flunkies at the *New York Times*—that all would be green again once sports utility vehicles (SUVs) and pickups were purged from the earth) was driven into news-network limbo. If for no other reason than that, we owe a small debt of gratitude to the Grope from Hope.

Green Media Scold Truck-Driving Public

But let us not revel in victory. At some point, Bubba's debasement of the presidency will become old news and the media's attention may once again shift to greener pastures, as it were, and a resumption of the global-warming cant. Sadly, the automobile industry has managed to react to this threat in what one of its best-known executives (who chooses here to remain anonymous) refers to as a "full Munich mode." Citing the mad rush at the 1998 North American International Automobile Show in Detroit to appear more like the Sierra Club than the Big Three (or Big 10, counting the major importers), he recently marveled at the sight of the industry "tripping all over ourselves to appear to be greener than the next guy."

Although such a move was celebrated by the media elite, it flies in the face of reality in the marketplace, where despite the screeches of outrage from the pundits, Americans are buying the hated SUVs, pickups, and vans at a furious rate.

These conveyances now account for about 47 percent of all passenger vehicles sold, up from a relatively paltry 20 percent in 1980. While editorialists at the *New York Times* clucked in shock over revelations that large vehicles like SUVs tend to fare better in crashes with smaller economy cars (to them, an amazing confirmation of 300-year-old Newtonian physics) and the greens swooned over their alleged gas guzzling, consumers flocked into showrooms to buy ever more powerful, more elaborate, more outsized Lincoln Navigators, Chevrolet Suburbans, Lexus LX470s, Ford Expeditions, Range Rovers, and AM General Hummers. As long as the economy remains strong and gasoline cheaper than Perrier water, the boom is likely to continue, with Cadillac soon to join the market.

Demonizing Automakers

Regardless of the fact that the new-generation SUVs are more fuel efficient than many of the main-line sedans they replaced and the public is speaking with its pocket-book, the campaign by the environmentalists to eliminate them remains as strong as ever. The industry can posture publicly

---

More Deaths from Smaller Cars

Climate change policies will have a lethal effect on people. They will kill more people through raising the federal Corporate Average Fuel Economy (CAFE) mandate for cars from 27.5 mpg to 45 mpg—a proposal pushed by several environmental groups. [Former] President Clinton recently one-upped this by promising to triple auto fuel efficiency over the next few years. But the human cost of CAFE is already too high—CAFE causes manufacturers to downsize cars in size and weight to meet the federal standard for their fleets, and smaller cars are much less safe than large cars in crashes.

According to a 1989 Harvard-Brookings study by Bob Crandall and John Graham, the current CAFE standard causes nearly 2,000 to 4,000 additional traffic deaths per year. If the standard were raised to 40 mpg, a 1992 study by Graham estimated, there would be an even greater increase in highway deaths—resulting in a total of 3,800 to 5,800 fatalities each year. *Each day*, from 10 to 16 people would die unnecessarily.

Frances B. Smith, *Consumer Alert*, September 1997.

---

about all manner of miraculous alternative energy sources—for example, fuel cells, electrics, solar, diesel electrics—but there appears little doubt that based on cost, efficiency, and ever-decreasing emissions, the gasoline-powered, internal-combustion engine will remain the powerplant of choice for the foreseeable future. At this point, other alternatives make no economic sense, period.

That said, nothing will deter the zealots. Not only are they convinced by the openly dubious science regarding global warming and the glories of electric cars, but they have also managed to delude the public that the industry is dominated by churlish profiteers bent on destroying the environment. As a *Detroit News* editorial noted recently, "The automakers' dilemma is real. A generation of Nader-ism has taught the public to doubt any claim by a private-sector company, particularly the Big Three." In the words of Jonathan Adler of the Competitive Enterprise Institute, it is a case of "Exxon's profits vs. Mother Teresa."

Assaulting Drivers with a Flawed Theory

Realism is doomed, and no one bothers to note that of the one-degree rise in temperature during the twentieth century, three-quarters of that increase took place before 1940—prior to mass industrialization and the alleged depredations of carbon dioxide. Totally ignored are findings from satellite data by the University of Alabama-Huntsville that the planet actually cooled slightly in 1997. In the midst of the hysteria about SUVs and trucks, one tends to forget a key finding by James Johnston of the American Enterprise Institute in his excellent book, *Driving America: Your Car, Your Government, Your Choice*. After analyzing government data on all sources of carbon-dioxide emissions that are the supposed culprit in global warming, Johnston discovered this: If all gasoline-burning vehicles—cars, SUVs, pickups, motorcycles, etc.— were removed from American highways, worldwide CO_2 emissions would be reduced by eighteen one-hundredths of one percent and the total emissions from man-made sources would drop by four percent. Johnston claims that motor vehicles contribute an even smaller percentage of water vapor and that "further reduction in the size and weight of motor vehicles to increase fuel economy would have no effect on global climate but would further erode the safety provided by larger, heavier vehicles." So much for the law of unintended consequences.

Sadly, none of these hard data, not to mention the largely unpublicized disagreement over the skewed conclusions of the Intergovernmental Panel on Climate Change (IPCC) will deter the lunatics. (Half the panel did not support the alarmist final report.) Yet the critics managed to seize the high ground among the media elite—much as Nader and his idiotic ravings did 30 years ago—and continue to promulgate their nonsense at will while serious, scientifically qualified skeptics are muted. Dr. Vincent Gray, a New Zealand–based scientist who headed the IPCC's review panel recently wrote, "Inaccurate data are preferred to accurate data, because the former fits a flawed theory."

Be warned. The assault on the automobile has only been delayed, not turned back. Once the nonsense in Washington involving, sadly, the rise of cynicism and the unraveling of the nation's moral fiber by a vapid but charming poltroon peters out, the war will resume, with more cries to increase Corporate Average Fuel Economy standards (easily done by executive order) to more pork-barreling mass-transit projects to the computerized Big Brother control of vehicle use.

> *"The fact is, U.S. nuclear power plants have reduced carbon dioxide emissions by more than 2 billion tons since 1973, more than any other energy source."*

Nuclear Power Is a Solution to Global Warming

Mary L. Walker

Mary L. Walker describes how a revival of the nuclear power industry could provide the United States with an important non–fossil fuel energy source, contributing to the reduction of greenhouse gas emissions as agreed upon by the Clinton Administration in the Kyoto Protocol. In Walker's opinion, a new generation of nuclear power plants can be built utilizing safer technology and presenting minimal risk to public safety and the environment. Walker was assistant secretary for environment, safety, and health at the U.S. Department of Energy from 1985 to 1988.

As you read, consider the following questions:

1. According to John Holdren as cited by the author, what is needed to meet the target for reducing greenhouse gas emissions?
2. Why does Walker believe that new nuclear plants will be cheaper to construct and operate?
3. In Walker's opinion, what issues are slowing the development of a permanent nuclear waste repository in Nevada?

Reprinted, with permission, from "Using Nuclear Power to Counter Global Warming," by Mary L. Walker, *The San Diego Union-Tribune*, September 16, 1999. Revised by the author, July 24, 2001.

Where will the world find leadership and technical expertise for the kind of global effort to curb environmental pollution that lies ahead in the new century? It can and should come from the United States.

Ecologist Rachel Carson first suggested more than two decades ago that if instead of allowing the indiscriminate use of DDT and other harmful pesticides Americans worked together to preserve the environment, we would add not only to our own well-being but to that of the planet as well. Policymakers responded to those concerns.

Increasing Demand for Limited Power

Today there is a huge new concern—the danger of global warming argued to be caused from the buildup of greenhouse emissions in the atmosphere. The seven warmest years on record, worldwide, have all been since 1990. Though the final verdict is out, one reason for the growing debate over curbing carbon dioxide emissions—the idea that something must be done to reduce the use of oil, coal, and natural gas—is that it has been difficult to discuss the issue objectively.

Inasmuch as power plants account for one-third of the greenhouse emissions that are linked to global warming, it is troubling that deregulation of the electric power industry has opened up a Pandora's box in California and other states. Deregulation has made it difficult to develop a coherent public policy that assures the need to avert future power failures will be met by using electricity more efficiently and building plants fueled by renewable energy sources and emission-free nuclear power. Unfortunately, California faces potential power shortages today because utilities stopped investing in new plants in the early 1990s.

The lesson we must take from inadequate power supplies is that our electricity supply system is critical to our economic and environmental well-being. In California, the demand for electricity has grown 15 percent since 1992, pushing utilities to the limit. Some 60 percent of the state's fossil-fuel plants are 30 years old—and cannot be pushed as hard as when they were built. So far, we have been able to stave off serious power shortages by using electricity purchased from other states.

However, increasing electrification of our economy means that we will need more power, not less.

Lowering Emissions with Nuclear Power

Electrical companies are proposing to build 20 new power plants in California that could add 13,500 megawatts of capacity. But almost all of the power plants would burn natural gas. Though it's cleaner than oil or coal, natural gas burned in ever-increasing volumes will make it harder to reduce smog and could trigger stricter controls for businesses and residents, since emissions of smog-producing nitrogen oxides would have to be offset by reductions elsewhere. Nitrogen oxides have been tied to respiratory problems in elderly people and children. And power plants fired with natural gas add to the atmosphere's greenhouse emissions.

This dilemma has led some policymakers—notably John Holdren, a professor of environmental polity at Harvard University—to advocate the revival of nuclear power. Though hardly a friend of the nuclear industry, Holdren—who played a key role years ago in shaping the Sierra Club's energy position—maintains there is little choice in the matter. He notes there are problems of one sort or another with every non-carbon energy source—whether solar, wind, biomass, geothermal, hydro or nuclear. He says we will need them all, as well as more energy-efficient technologies, if we hope to meet the target for reducing greenhouse emissions that was agreed upon in the Global Warming Treaty.

The fact is, U.S. nuclear power plants have reduced carbon dioxide emissions by more than 2 billion tons since 1973, more than any other energy source. And the plants do not produce any of the pollutants that cause acid rain and smog.

Improved Safety from Advanced Plants

This makes possible the inevitable—and quite desirable—next generation of nuclear plants. These advanced power plants which are mid-size reactors in the 600-megawatt range rely primarily on passive safety systems that would make future nuclear units even safer than existing units. In the extremely remote possibility that anything goes wrong, the plants will automatically shut down and self-cool.

Because the advanced plants would be built from standardized designs, patterned after the compact reactors in the nuclear Navy, their construction and operating costs will be substantially lower than today's plants.

Liabilities of Energy "Business as Usual"

If the United States and the world continue to careen down the path of energy "business as usual"—content to assume that tomorrow's energy-supply system can be much like today's, except larger—we will pay for this complacency with higher energy costs and lower energy security, slower economic growth, excessive environmental impacts, and increased international tensions in the decades ahead. Specifically:

- The United States—and many other countries as well—will be increasingly dependent on oil from the Middle East . . . and correspondingly vulnerable to externally imposed price hikes and supply disruptions. The potential for armed conflict over access to oil supplies will grow. . . .

- Disruption of global climate from the atmospheric buildup of heat-trapping gases—above all carbon dioxide from fossil-fuel combustion—will become the dominant environmental problem of the 21st century, imperiling the productivity of farms, forests, and fisheries, rendering many of the world's cities increasingly unlivable in summer, putting coastal property and wetlands at risk from rising sea level, and imposing a panoply of other adverse impacts on human health, property, and ecosystems.

- Economic growth will be curtailed and economic aspirations frustrated—especially in the developing countries but also to some extent in the industrialized ones—by constraints on the growth of energy supply imposed by rising monetary and environmental costs and by disputes over energy choices and facility siting.

John P. Holdren Testimony for the Subcommittee on Energy and Environment Committee on Science, U.S. House of Representatives, July 25, 2000.

The Nuclear Regulatory Commission has already approved designs for the advanced reactors. It also has adopted changes in the licensing system to give citizens greater input in the hearing process.

The only thing missing is the money to construct the first compact plant. An obvious solution is for private industry and government to enter into a cost-sharing partnership.

This partnership between government and industry would be able to demonstrate that a safe, compact nuclear plant can be built on time and within budget. Importantly, we would be taking action to meet overall energy security and global warming goals.

Those who cite the nuclear waste problem as an insurmountable barrier ignore the fact that the issues slowing development of a permanent waste repository at Yucca Mountain in Nevada are largely related to states rights and are political, not technical. The Nuclear Regulatory Commission continues to move ahead on licensing requirements for Yucca Mountain, while the Department of Energy completes development of the repository.

Legitimate concern about government interference in the marketplace should not be permitted to muddle what remains the essential point: the United States needs to expand the use of nuclear power in order to meet the goals for reducing air pollution and greenhouse emissions. If we are to have a reliable and non-polluting power system, the time to remove obstacles to nuclear power's future is now. Otherwise, we will be moving into the 21st century with one carbon fuel to burn: natural gas. That's a risky strategy for California and the nation.

"Tackling climate change through the development of nuclear power is both expensive and just swaps one serious problem for another."

Nuclear Power Is Not a Solution to Global Warming

Friends of the Earth Scotland

Friends of the Earth Scotland is an organization campaigning to bring about renewable energy, better public transportation, and a cleaner environment. In the following viewpoint, the organization argues that nuclear power is not a practical alternative to fossil-fuel energy sources in attempting to reduce the greenhouse gas emissions associated with global warming. According to the authors, nuclear power is not cost effective in reducing carbon dioxide emissions, poses serious problems to environmental safety, and contributes carbon dioxide to the atmosphere through uranium mining and plant construction.

As you read, consider the following questions:

1. In comparison to renewable energy, how much more carbon dioxide per unit of energy produced does nuclear power release into the atmosphere, in the authors' opinion?
2. According to the authors, where did a serious nuclear accident occur in 1986?
3. What is the unavoidable legacy of nuclear power, according to Friends of the Earth?

Reprinted, with permission, from "Nuclear Power Is No Solution to Climate Change: Exposing the Myths," January 1998 Friends of the Earth Scotland briefing paper found at www.nirs.org/KYOTONUC.html.

The nuclear industry is hoping that concern over climate change will result in support for nuclear power. However, even solely on the grounds of economic criteria it offers poor value for money in displacing fossil fuel plants. Further, with its high cost, long construction time, high environmental risk and problems resulting from waste management, it is clear that nuclear power does not offer a viable solution to climate change. Rather a mixture of energy efficiency and renewable energy offers a quicker, more realistic and sustainable approach to reducing carbon dioxide (CO_2) emissions.

Exposing the Myths 1: Nuclear Power Is Economical and Cost Effective

The full costs of nuclear power have been seriously underestimated by all countries which have the technology, and it is only recently that the true costs have begun to come to light. The hidden costs of waste disposal, decommissioning and provision for accidents have never been adequately accounted for, resulting in a massive drain upon economies. This drain will continue for many years to come as the expensive and dangerous task of nuclear decommissioning gets underway.

Privatisation and liberalisation of the market in the United Kingdom (UK), has led to the true costs of nuclear power being exposed. It has become clear that nuclear power cannot exist in a competitive energy market without significant subsidy from the UK Government. This process is now being followed around the world with investors being unwilling to accept the high cost and risks associated with nuclear power. Moreover, if fully comprehensive insurance was required to cover all of the risks of nuclear accidents, the cost of electricity from nuclear power would increase many times from the present level.

Reactor decommissioning costs also remain a major uncertainty. In the UK, for example, the cost of dealing with the unwanted debris of the nuclear industry is officially estimated at about US$70 billion. Of this, just US$22 billion is covered in secure funding arrangements, with the remaining US$48 billion (almost 70%) likely to be paid for by taxpayers. The nuclear industry's claim that, "In most countries, the full costs of waste management and plant decommissioning will be funded from reserves accumulated from current revenues" is clearly untrue.

Countries, particularly in Central and Eastern Europe, are continuing to build new nuclear plants even though it has been shown that investment in energy efficiency measures is the quickest and safest way to tackle their energy crises. For example, the nuclear power plants proposed to replace the remaining reactors at Chernobyl have consistently been shown not to be the least-cost option.

Also, in terms of cost-effectiveness in reducing CO_2 emissions, nuclear power fairs very poorly. In 1995, after a year-long, exhaustive review of the case for nuclear power, the UK Government concluded that nuclear power is one of the least cost-effective ways in which to cut CO_2 emissions. In the USA improving electricity efficiency is nearly seven times more cost effective than nuclear power for obtaining emissions reductions.

Exposing the Myths 2: Nuclear Power Does Not Produce CO_2

Nuclear power is not greenhouse friendly. While electricity generated from nuclear power entails no direct emissions of CO_2, the nuclear fuel cycle does release CO_2 during mining, fuel enrichment and plant construction. Uranium mining is one of the most CO_2 intensive industrial operations and as demand for uranium grows CO_2 emissions are expected to rise as core grades decline.

According to calculations by the Öko-Institute, 34 grams of CO_2 are emitted per generated kilowatt-hour (kWh) in Germany. The results from other international research studies show much higher figures—up to 60 grams of CO_2 per kWh. In total, a nuclear power station of standard size (1,250 megawatts operating at 6,500 hours/annum) indirectly emits between 376,000 million tonnes (Germany) and 1,300,000 million tonnes (other countries) of CO_2 per year. In comparison to renewable energy, nuclear power releases 4–5 times more CO_2 per unit of energy produced taking account of the whole fuel cycle.

Also, with its long development time a nuclear power programme offers no short-term possibility for reducing CO_2 emissions.

Exposing the Myths 3: Nuclear Power Is Safe

Problems of security, safety and environmental impact have been perennial issues for the nuclear industry. Many countries have decided against the development of nuclear power on these grounds, but radioactive contamination is no respector of national borders and nuclear power plants threaten the health and well-being of all surrounding nations and environments. There are also the very serious problems of nuclear proliferation and trafficking.

The United Nations' Intergovernmental Panel on Climate Change (IPCC) view is that if nuclear power were to be used extensively to tackle climate change, "The security threat . . . would be colossal".

Enormous Costs

The external costs of nuclear power include the cost of environmental damage, the effect on human health and society following an accident, damage to human health and the environment during routine operation of nuclear facilities and also long term problems associated with nuclear waste and decommissioning of nuclear facilities. "Externalities" that lend themselves to monetary quantification include economic effects, employment, environment, environmental impacts, health effects and government subsidies.

When such quantifiable social costs are added to the core price of electricity, the total costs of nuclear power are extremely high. Nuclear power no longer stays competitive against the latest generation of renewable energy.

Greenpeace, "Nuclear Energy No Solution to Climate Change: A Background Paper," n.d.

Just one month after *The Economist*, a British magazine, had declared in its lead article that the technology was "as safe as a chocolate factory" (1986), there followed a catastrophic nuclear accident at Chernobyl. The accident caused an immediate threat to the lives of 130,000 people living within a 30 kilometre radius who had to be evacuated (and who have been permanently relocated) and 300–400 million people in 15 nations were put at risk of radiation exposure. Forecasts of additional cancer deaths attributable to the Chernobyl accident range from 5,000 to 75,000 and beyond. The nuclear industry argues that the problems in the former Soviet Union are different to those in developed countries, but the United States itself had a serious accident at Three Mile Island in 1979. Whilst the new European Pressurised Reactor and the fusion programmes are being promoted as offering safer operation, no form of nuclear power technology is totally without risk of a major accident. With public opinion strongly set against nuclear power, it would be far better to invest in renewable forms of energy which have widespread public support. The development of new nuclear technology would mean spending huge amounts of money going down another nuclear road, with the prospect of finding the same type of problems and public opposition.

Recent in-depth studies in the United States challenge the claim that exposure to low-level doses of radiation is safe. The health and safety of employees, local communities and the contamination of the environment are genuine risks. A study (completed August 1997) funded by the US National Institute for Occupational Safety and Health of the Centers for Disease Control and Prevention, examined the health and mortality of 14,095 workers from the Oak Ridge National Laboratory. The study found "strong evidence of a positive association between low-level radiation and cancer mortality". As of 1990, 26.9% of deaths were due to cancer.

The exposure risk to workers in the uranium mining industry is also great.

Exposing the Myths 4: Nuclear Power Is Sustainable

Nuclear power plants produce extremely long-lived toxic wastes, for which there is no safe means of disposal. The only independent scrutiny of a UK Government waste management safety case led to the cancellation of the proposed test site for nuclear waste disposal. As disposal is not scientifically credible, there is no option other than interim storage of radioactive wastes. This means that the legacy of radioactive wastes will have to be passed on to the next generation. Producing long-lived radioactive wastes, with no solution for their disposal, leaving a deadly legacy for many future generations to come is contrary to the principle of sustainability. . . .

In 1976 the UK Royal Commission on Environmental Pollution warned that it is "irresponsible and morally wrong to commit future generations to the consequences of fission power on a massive scale unless it has been demonstrated beyond reasonable doubt that at least one method exists for the safe isolation of these wastes for the indefinite future". Over twenty years on, still no such method has been found. Nuclear waste management policies are in disarray and there is growing public opposition to the transport and storage of nuclear waste. . . .

Under no circumstances can nuclear power be considered to be sustainable.

With the virtual demise of the Fast Breeder research programme and no foreseeable commercial development of fusion reactors, the belief that nuclear power can supply an endless source of energy is fast disappearing. The Japanese Monju Fast Breeder reactor has been inactive since a serious accident in December 1995, whilst the French Superphoenix and the breeder reactor programmes in the UK have been permanently closed.

Diminishing uranium supplies and the failure of the breeder reactor programmes mean that nuclear power will not be able to make a long-term contribution to meeting the world's energy needs.

Exposing the Myths 5: Nuclear Power Makes a Vital Contribution to Energy Supply

The assertion by the nuclear industry that, "It is essential that nuclear generating capacity is maintained if emissions from power generation are to be successfully limited over the next 10 to 15 years and beyond" is fundamentally untrue. Emissions can be cut without building more nuclear power plants. In October 1997, the US Department of Energy released a report in which they concluded that the US could cut CO_2 emissions to 1990 levels by 2010 with no net cost to the economy. Shell has forecast that renewables could meet up to 50% of the world's energy demand by 2060. Nuclear power only supplies 17% of world electricity supply at present.

Nuclear power is seeing its role in the world's energy mix diminish. Since 1986, according to the International Atomic Energy Agency, only three nuclear power stations have been ordered annually. In Europe fourteen out of fifteen European nations have no plans to develop nuclear power; the majority of the countries within the European Union have "little desire to launch, or to re-invigorate, nuclear power programs"; and nearly half of the EU countries are nuclear free and others are planning to decrease or phase out nuclear power completely. It is clear that the vast sums of money being spent on research and development and on subsidising the industry are in total disproportion to the contribution nuclear power is likely to make to Europe's energy supply in the coming decades.

With a limited amount of funding available for research and development, reallocation of funds from nuclear power and towards renewable energy and energy efficiency would reduce the costs of these technologies, making them even more competitive. However, funds are still being wasted on nuclear power programmes, which are opposed by most people, are more expensive than other alternatives and require a long development time.

It is a myth that "Nuclear power is the only fully developed non–fossil fuel electricity generating option with the potential for large-scale expansion". Nuclear power plants take 10 years to build. Over the next 12 years the European Union is aiming for 10,000 megawatts of wind power and 10,000 megawatts of biomass to be developed. This is just a part of the solution and is equivalent to about 15 nuclear power plants. . . .

No Nuclear Solution to Climate Change

Under no circumstances can nuclear power be considered to be a solution to climate change:

• It is one of the most expensive ways to reduce carbon dioxide emissions.

- The nuclear industry does contribute to carbon dioxide emissions.
- No proven strategies exist for the permanent safe storage of nuclear waste.
- Nuclear power poses a very real health risk.
- Nuclear power is uneconomic, unsustainable and unsafe.

Climate change is a serious problem which needs to be tackled in a way which safeguards the future for generations to come. Tackling climate change through the development of nuclear power is both expensive and just swaps one serious problem for another. Nuclear power cannot be considered to be a "clean source of electricity" [as the nuclear power industry claims].

The nuclear industry is hoping to use the Climate Change negotiations to save itself, because the economics of nuclear power has meant a rapid decline in the industry's fortunes. This is a desperate attempt to generate business from the misfortune of the problems we all now face.

> "Better forestry could enable the United
> States to meet its [Kyoto] treaty
> commitment instead of . . . shutting down
> coal plants and jacking up gasoline taxes."

Planting Trees Can Help Combat Global Warming

Thomas M. Bonnicksen

In the following viewpoint, Thomas M. Bonnicksen maintains that the United States can reduce carbon dioxide emissions as required under the Kyoto Protocol, a global climate
agreement, by improving the health of its forests and planting trees on fallow land. According to Bonnicksen, forests
act as "carbon sinks" and absorb carbon dioxide as they
grow. The capacity of trees to soak up carbon dioxide could
cut in half the fossil-fuel emissions reductions required of
the United States, in the author's opinion. Bonnicksen is a
professor of forest science at Texas A&M University.

As you read, consider the following questions:
1. In the author's opinion, how might the carbon-sink effect
   of planting trees reduce pressure on American
   companies?
2. What evidence included in recent studies has
   demonstrated the carbon-sink strategy to be effective
   against global warming, according to the author?
3. In Bonnicksen's opinion, how can forest health and
   productivity be improved?

Reprinted, with permission, from "Forests Can Give Us Breathing Room on
Kyoto," by Thomas M. Bonnicksen, *Houston Chronicle*, November 15, 2000.

If anything should make skeptics consider the possibility that there is something new or promising about the treaty to curb global warming, it is the growing recognition that the world's forests and agricultural land hold the key to any final agreement for stabilizing the climate. The focus on forests and farmland could have salutary results for Texas.

A Natural Solution to Reducing Emissions

Until now, international talks aimed at carrying out the 1997 Kyoto Protocol on global warming have centered on cutting emissions from power plants and motor vehicles. However, the United States proposed treaty negotiations in November 2000 that the natural capacity of America's forests and farmland to absorb carbon dioxide should give the nation substantial credit in meeting its requirements under the Kyoto treaty.

Experts calculate that the so-called carbon-sink effect of America's forests and farmland could cut by as much as half the carbon dioxide reductions needed to comply with the treaty. That would greatly reduce pressure on American companies, since it promises to keep the cost of fighting global warming down, especially in Texas and other energy-rich states that produce and burn large amounts of oil, coal and natural gas.

Scientists have known for decades that trees and other plants absorb carbon dioxide as they grow. However, the carbon-sink strategy grows out of recent evidence that forests soak up much more carbon than previously thought. Earlier studies had neglected to include the huge amounts of carbon stored in peat and other organic matter in soils—now estimated to account for two-thirds of the total sequestered.

These are not small numbers. According to government projections, U.S. emissions will probably exceed 2.1 billion tons of carbon by 2008. The Kyoto treaty requires the United States to reduce that to 1.5 billion tons. If the United States gets credit for the carbon-storing capabilities of forests and farmland the way they are currently managed, they are worth roughly 300 million tons a year.

However, an all-out effort to improve the health of U.S. forests and plant more trees on marginal land could more than double the carbon sink to 600 million tons. In other words, better forestry could enable the United States to meet its treaty commitment instead of weighing draconian mandates such as shutting down coal plants and jacking up gasoline taxes.

Cost-Effective Carbon Storage

Converting fallow land to forests would be quite cost-effective compared to some regulatory proposals to cut greenhouse emissions. Under the Federal Conservation Reserve Program, an estimated 4 million to 5 million acres of marginal crop and pastureland once used for farming has been converted to timberland. With appropriate incentives to landowners, more than 100 million acres of marginal land considered biologically suitable for trees (an area the size of California) could be reforested. More trees also mean healthier air and less need for air conditioning, if trees are planted near buildings.

Another way to increase carbon storage is to improve forest health and productivity. A combination of logging, thinning and careful use of prescribed fires to remove underbrush helps protect forests from wildfires, especially in high-risk areas where trees have been weakened by disease and insects. Since wildfires emit huge amounts of carbon dioxide, a vigorous national program aimed at making forests healthier must be a top priority.

It's too bad that national environmental groups are too caught up in anti-logging sentiment to recognize the value of better forest management.

Make no mistake; this is not to imply that all logging is beneficial to the environment. The U.S. forest products industry could send a message around the world to preserve tropical rain forests by helping to sustain and restore dwindling ancient forests in this country. Such efforts by industry could blunt the argument of developing nations that the United States is more interested in preserving their natural resources than it is its own.

Climate change affects us all. Instead of being discounted, forests are an essential part of the world community's efforts to combat global warming. We can succeed if our country, spurred by the Kyoto treaty, embarks on a course of wise, sustainable action to make better use of trees.

"Creating forests to soak up the greenhouse gas carbon dioxide . . . [is] not a panacea for global warming."

Planting Trees Cannot Substitute for Reducing Fossil-Fuel Emissions

Fred Pearce

Fred Pearce, an environmental journalist, argues in the following viewpoint that planting trees to soak up carbon dioxide is not an effective alternative to reducing emissions from the burning of fossil fuels. The United States and other industrialized countries have proposed allowing planned new forests called "carbon sinks" to count toward their total greenhouse gas reductions as required by the Kyoto Protocol, a global climate treaty. According to Pearce, new forests may actually accelerate global warming by returning carbon dioxide to the atmosphere.

As you read, consider the following questions:
1. What fraction of the total carbon dioxide emissions from human activity is absorbed by the world's forests, according to Pearce?
2. In the author's opinion, how long is the time lag between the absorption and outpouring of carbon dioxide from the world's forests?
3. How much carbon, according to Pearce, does the Norwegian forestry company Treefarms predict its forest will store by 2010?

Reprinted from "That Sinking Feeling," by Fred Pearce, *New Scientist*, October 23, 1999, with permission.

Not for the first time, the human race may be about to take a dangerous short cut. In October 1999, governments from around the world met in Bonn, the former German capital, to complete plans for creating forests to soak up the greenhouse gas carbon dioxide. They see planting trees as a partial alternative to cutting emissions of the gas from power stations and vehicle exhausts.

Carbon Sinks Versus Emissions Reductions

Unfortunately, the United Nations' (UN) Intergovernmental Panel on Climate Change (IPCC) put the finishing touches to a report that shows this strategy to be based on a dangerous delusion. In reality, say its scientists, planned new forests, called "carbon sinks", will swiftly become saturated with carbon and begin returning most of their carbon to the atmosphere, temporarily accelerating global warming. Peter Cox of the Hadley Centre, part of Britain's Meteorological Office, shares the UN panel's conclusions. "This is not something that may or may not happen as the world warms—it is more or less inevitable," he says. The result will be no overall reduction in carbon dioxide (CO_2) levels. Despite this, the US and other major CO_2 producers will cite the "sink" process as justification for their continuing tardiness in cutting CO_2 emissions. The US Environmental Protection Agency would not comment when approached by *New Scientist*. But other governments are concerned. Britain's environment department says: "The UK emphasises that the main action should be reducing actual emissions. Sinks are much less secure than carbon in fossil fuels left unburnt."

The discovery that forests are not a panacea for global warming only emerged after they were given a central role in the Kyoto Protocol, the treaty signed in 1997 by most of the world's governments in a bid to stem the greenhouse effect. "Just a couple of years ago, the issue of sink saturation was barely known," says Will Steffen of Sweden's Royal Academy of Sciences, who chairs the International Geosphere-Biosphere Programme (IGBP), which has pioneered research into the global carbon cycle. The first public warning came in the March 1999 issue of the IGBP newsletter. And in October 1999, the IPCC incorporated its analysis into a forthcoming report on land use change and forestry.

Pollution and Carbon Sinks

Each year, CO_2 emissions from human activity pour just over 6 billion tonnes of carbon into the atmosphere. Around a third is absorbed by the world's forests. The discovery of this large carbon sink encouraged policy makers to believe that CO_2 pollution could be cut by planting more trees. But now it seems the sink is a recent phenomenon, and a temporary one. In fact, the suggestion that planting trees means less atmospheric CO_2 ignores simple logic.

Before the large-scale development of industry, mature forests were in equilibrium with the atmosphere. Photosynthesis, the process that creates plant matter, absorbs CO_2 from the air. But trees also release CO_2 back into the air when plant matter breaks down the sugars they make during photosynthesis. This process is called respiration. Much the same happens in forest soils, which absorb carbon from trees and release CO_2 as microorganisms break down plant matter.

This equilibrium has been increasingly upset by the higher concentrations of CO_2 in the atmosphere from burning fossil fuels. Usually, the low level of CO_2 is what limits photosynthesis. Higher CO_2 levels promote "CO_2 fertilisation", accelerating both forest growth and the accumulation of carbon in forest soils. And as the forests grow faster, they absorb more CO_2, helping to stave off climate change—for a while.

Planting Forests May Contribute to Global Warming

Until recently, researchers had assumed that as long as CO_2 levels in the air went on rising, the forest sink would continue to grow. The IPCC's 1996 assessment concluded that forests would soak up around 290 billion tonnes of carbon next century, even without extra planting. But this now seems highly unlikely. Experts such as Bob Scholes of the South African government's research agency, Council of Scientific and Industrial Research, argue that CO_2 fertilisation may already have peaked and that respiration may be about to accelerate. Early in the 21st century, forests planted to protect the planet from global warming could be contributing to it.

How did researchers get it so wrong? Scholes, a leading light in the IGBP's Global Carbon Project, says that the confusion was caused by a time-lag. CO_2 fertilisation is an instantaneous process. But respiration increases in response to temperature rises triggered by the CO_2. That warming has a built-in delay of about fifty years, caused largely by the thermal inertia of the oceans. So the extra outpouring of CO_2 from the world's forests would not yet be apparent. "During this delay there is an apparent carbon sink," he says.

Don't Delay Fossil-Fuel Reductions

Fixing atmospheric carbon in biological systems by, for instance, planting trees, is no substitute for reducing the burning of fossil fuels. Fossil fuels release into the Earth-atmosphere system ancient carbon from the planet's crust, where it was locked safely away for millions of years. . . .

Forest programs should be separate from national obligations to reduce fossil fuel emissions. Since tree planting is comparatively cheap, it will be the economic choice to the extent allowed by the Kyoto Protocol. The extraction and release of ancient carbon will continue unabated and the mass commercialization of emerging carbon-free renewable energy sources will be delayed.

Rhys Roth, Atmospheric Alliance, April 6, 1998.

At the Hadley Centre, Cox has just finished modelling the likely future carbon cycle. He warns that we are on a "saturation curve", where extra CO_2 has an ever-smaller effect on plant growth. Respiration, on the other hand, continues to increase with temperature. Soil respiration in particular goes up exponentially with temperature, at least for a time, says Cox. So if CO_2 levels in the air continue to rise, fertilisation rates will flatten out while respiration rates soar. He predicts that by 2050, forests will have released much of what they have absorbed. The overall reduction in CO_2 levels will therefore have been small. "The timing is uncertain but we are pretty certain it will happen," he says. Wolfgang Cramer of the Potsdam Institute in Germany has recently reached a similar conclusion. Neither study is yet published. The effect of accelerated respiration on the atmosphere could be even more dangerous if, as predicted by some scientists, the heat and drought caused by global warming degrade tropical forests at the same time.

Abandoning Substantive Reductions

This isn't to say planting trees is in itself a bad thing. Whether they are absorbing or releasing the gas, they will always be keeping some CO_2 out of the atmosphere and providing other ecological benefits. But forests are an insecure way of storing carbon out of harm's way, says Steffen. The real danger, he says, arises when countries use plans to plant forests as a justification for not cutting their CO_2 emissions from burning fossil fuels. In October 1999 politicians and climate negotiators met in Bonn . . . to agree to rules for implementing the 1997 Kyoto Protocol, which give the planting of forest sinks equal value to emissions cuts as a way to meeting its targets.

Many countries, including the US, which produces about a quarter of the world's CO_2 emissions, are relying to a large degree on the supposed benefits of tree planting to meet their targets. And dozens of forestry companies are already planning to join the anticipated global market in certificated carbon sinks. Typical is a new project, announced by a Norwegian forestry company, to plant fast-growing pine and eucalyptus trees on 150 square kilometres of grassy plain in southwest Tanzania. The company, called Treefarms, promises that by 2010 the Kilombero Forest will store more than a million tonnes of carbon. But will it? Such claims are based on models of CO_2 accumulation that assume current rates of CO_2 fertilisation will continue. But Scholes believes the carbon sink will start to decline within the next few decades. This would make the certificates for carbon stored in forests such as Kilombero worthless.

Ultimately, says Steffen, we will only save the world from catastrophic climate change by cutting emissions. "New forests are temporary reservoirs that can buy valuable time to reduce industrial emissions, not permanent offsets to these emissions." But the Kyoto Protocol does not reflect that—and nor did October 1999's treaty negotiations. It could prove a devastating mistake.

"The carbon cycle has a very long equilibrium time," says Scholes. "The consequences of actions taken now will persist for many centuries."

Periodical Bibliography

The following articles have been selected to supplement the diverse views presented in this chapter. Addresses are provided for periodicals not indexed in the *Readers' Guide to Periodical Literature*, the *Alternative Press Index*, the *Social Sciences Index*, or the *Index to Legal Periodicals and Books*.

Gary S. Becker
"What Price Pollution? Leave That to a Global Market," *Business Week*, October 18, 1999.

Robin Cook
"Everything to Gain," *Our Planet*, vol. 9, no. 3, 1997. Available from PO Box 30552, Nairobi, Kenya.

Jennifer Couzin
"The Forest Still Burns," *U.S. News and World Report*, April 19, 1999.

John H. Cushman Jr.
"U.S. Signs Pact to Reduce Gases Tied to Warming," *New York Times*, November 13, 1998.

Gregg Easterbrook
"Get the Easy Greenhouse Gases First," *New York Times*, August 29, 2000.

Marsha Freeman
"Defeat the Kyoto Protocol," *21st Century Science & Technology*, Winter 1997/1998. Available from PO Box 16285, Washington, DC 20041.

Darren Goetze
"Triple Play," *Nucleus*, Spring 1999. Available from Two Brattle Square, Cambridge, MA 02238.

Jennifer G. Hickey
"Flaky Climate Data Will Cost U.S. Dough," *Insight on the News*, December 15, 1997. Available from 3600 New York Ave. NE, Washington, DC 20002.

Donald Hodel and Fred Smith
"Climate Is Not Right for Global-Warming Treaty," *Insight on the News*, December 29, 1997.

Jack Kemp
"Yet Another Attack on the Economy," *San Diego Union-Tribune*, March 6, 2001. Available from PO Box 120191, San Diego, CA 92112-0191.

Los Angeles Times
"Warming Issue Persists," January 27, 2001. Available from 202 W First St., Los Angeles, CA 90012.

Jim Motavalli
"Why Detroit's Going Green," *Sierra*, July/August 1999.

New York Times	"A Global Warning to Mr. Bush," February 26, 2001.
Andrew C. Revkin	"Treaty Talks Fail to Find Consensus in Global Warming," *New York Times*, November 26, 2000.
Ernst D. Schulze	"Managing Forests After Kyoto," *Science*, September 22, 2000.
Chuck Sudetic	"The Good News About Global Warming," *Rolling Stone*, October 12, 2000.
Dick Thompson	"What Global Warming?" *Time*, June 21, 1999.

For Further Discussion

Chapter 1

1. Ross Gelbspan asserts that media coverage of the global warming debate gives equal weight to both skeptics and supporters of global warming. S. Fred Singer maintains the opposite—that the media overwhelmingly cover global warming as a serious problem and attempt to discredit critics by identifying them with the oil industry. What evidence does each give in support of his argument? Who makes the more convincing case against the media? Cite examples from the text to support your answer.

2. S. Fred Singer writes that the global warming debate has become less a question of science than of politics, with left-wing, Democratic proponents challenging right-wing, Republican skeptics. Does the politicized debate affect your assessment of the scientific research that both sides use to defend their positions on global warming? Explain your answer using examples from the James K. Glassman and National Assessment Synthesis Team viewpoints.

3. The Union of Concerned Scientists (UCS) acknowledges that scientists cannot know for certain the exact impact of global warming on the earth's climate, although computer climate models suggest that the planet is getting warmer. Gene Barth claims that these climate models are inaccurate and exaggerate the extent of global warming. Is it "unrealistic and unnecessary," as the UCS argues, to expect scientific certainty before taking action to address the threat of global warming? Why or why not?

Chapter 2

1. Robert T. Watson believes that the earth's climate has been relatively stable over the last 10,000 years but has recently undergone a dramatic rise in temperature due to human activity. What evidence from earth's geologic past does John Carlisle present to refute the idea of a stable climate?

2. The majority of scientific experts, according to Robert T. Watson, believe that human-induced global warming is inevitable but cannot be reversed quickly, if at all, due to the long life of greenhouse gases in the atmosphere. Do these facts undermine Watson's argument that reductions in greenhouse gas emissions should be undertaken? Explain your answer.

3. John L. Daly is in agreement with Judith Lean and David Rind that changes in the sun's radiation most likely caused the Little Ice Age that occurred from 1610 to 1800. What reasons do Lean and Rind give for the weakening influence of the sun on the earth's rising temperatures in the years following the Little Ice Age? How does their reasoning contradict Daly?

Chapter 3

1. Gar Smith contends that global warming will increase the occurrence of extreme weather events like hurricanes, floods, and droughts. How does Dennis T. Avery address this contention? Would you prefer to live in a warmer or colder climate? Support your answer with examples from these viewpoints.

2. Paul Kingsnorth expresses alarm over global warming's effect on the spread of malaria to northern cities like New York. What evidence does Thomas Gale Moore provide in his assertion that global warming has little to do with the spread of malaria to northern regions? Are his arguments convincing? Why or why not?

3. Richard D. Terry argues that measuring sea level rise is nearly impossible and leaves researchers with flawed data. Stuart R. Gaffin contends that there is compelling data documenting a significantly accelerating sea level rise over the past 100 years. Which viewpoint makes the more convincing argument? Why?

Chapter 4

1. The Clinton administration believed that agreeing to the emissions reductions of the Kyoto Protocol is like taking out a reasonably priced insurance policy on the environment. After reading George Melloan's viewpoint, does the Protocol fit the administration's description? Why or why not?

2. Jane Holtz Kay wants to raise the tax on gasoline to curtail automobile use. Would this taxation have a positive or negative impact on the United States? Explain. Do you agree with *Car and Driver* that government regulation of automobile use impinges on personal freedom? Explain.

3. Mary L. Walker claims that nuclear power plants have reduced carbon dioxide emissions by more than 2 billion tons since 1973. Given this significant reduction, do you believe that the benefits of reviving nuclear power outweigh its environmental dangers, as outlined by the Friends of the Earth Scotland? Explain your answer with examples from the viewpoints.

Organizations to Contact

The editors have compiled the following list of organizations concerned with the issues debated in this book. The descriptions are derived from materials provided by the organizations. All have publications or information available for interested readers. The list was compiled on the date of publication of the present volume; the information provided here may change. Be aware that many organizations take several weeks or longer to respond to inquiries, so allow as much time as possible.

Center for Global Change Science (CGCS)
77 Massachusetts Ave., MIT 54-1312, Cambridge, MA 02139
(617) 253-4902 • fax: (617) 253-0354
e-mail: cgcs@mit.edu • website: http://web.mit.edu/cgcs/www/
CGCS at the Massachusetts Institute of Technology addresses long-standing scientific problems that impede accurate predictions for changes in the global environment. The long-term goal of CGCS is to accurately predict environmental changes by utilizing scientific theory and observations to understand the basic processes and mechanisms controlling the global environment. The center publishes and distributes a Report Series of papers intended to communicate new results and provide reviews and commentaries on the subject of global climate change.

Climate Solutions
610 Fourth Avenue E, Olympia, WA 98501
(360) 352-1763 • fax: (360) 943-4977
e-mail: info@climatesolutions.org
website: http://climatesolutions.org
Climate Solutions' mission is to stop global warming at the earliest point possible by helping the northwest region of the United States develop practical and profitable solutions. It focuses on job creation, economic development, and environmental protection. Climate Solutions publishes reports covering the impact of global warming, including *Global Warming Is Here: The Scientific Evidence*, available from its website.

Competitive Enterprise Institute (CEI)
1001 Connecticut Ave. NW, Suite 1250, Washington, DC 20036
(202) 331-1010 • fax: (202) 331-0640
e-mail: info@cei.org • website: www.cei.org/ceimain.asp
CEI is a nonprofit public policy organization dedicated to the principles of free enterprise and limited government. CEI encour-

ages the use of private incentives and property rights to protect the environment. Instead of government regulation, it advocates removing governmental barriers and establishing private sector responsibility for the environment. CEI's publications include the monthly newsletter *CEI Update*, *On Point* policy briefs, and the books *The True State of the Planet* and *Earth Report 2000*.

The George C. Marshall Institute

1730 K St. NW, Suite 905, Washington, DC 20006
(202) 296-9655 • fax: (202) 296-9714
e-mail: info@marshall.org • website: www.marshall.org

The institute is a nonprofit research group that provides scientific and technical advice and promotes scientific literacy on matters that have an impact on public policy. It is dedicated to providing policy makers and the public with rigorous, clearly written, and unbiased technical analyses of public policies. The institute publishes several studies on global warming including *A Scientific Discussion of Climate Change* and *Are Human Activities Causing Global Warming?*

Global Warming International Center (GWIC)

22W381 75th Street, Naperville, IL 60565
(630) 910-1551 • fax: (630) 910-1561
website: www.globalwarming.net

GWIC is an international body that disseminates information on global warming science and policy to governmental and non-governmental organizations and industries in more than 120 countries. The center sponsors unbiased research supporting the understanding of global warming and its mitigation. GWIC publishes the quarterly newsletter *World Resource Review*.

Goddard Institute for Space Studies

2880 Broadway, New York, NY 10025
(212) 678-5641
e-mail: emichaud@giss.nasa.gov • website: www.giss.nasa.gov

GISS, a division of the National Aeronautics and Space Administration (NASA), is an interdisciplinary research initiative addressing natural and man-made changes in the environment that affect the habitability of the planet. A key objective of GISS research is the prediction of atmospheric and climate changes in the twenty-first century. The institute publishes numerous research papers on global warming that are available from its website.

The Heartland Institute
19 South LaSalle #903, Chicago, IL 60603
(312) 377-4000 • fax: (312) 377-5000
e-mail: think@heartland.org • website: www.heartland.org

The Heartland Institute is a nonprofit public policy research organization that provides research and commentary to elected officials, journalists, and its members. The institute conducts policy studies on global warming and other environmental issues. It publishes the monthly *Environment & Climate News* newspaper and the book *Eco-Sanity: A Common-Sense Guide to Environmentalism*.

The Heritage Foundation
214 Massachusetts Ave. NE, Washington, DC 20002
(202) 546-4400 • fax: (202) 546-8328
e-mail: info@heritage.org • website: www.heritage.org

The Heritage Foundation is a conservative think tank that supports free enterprise and limited government in environmental matters. Its publications, such as the quarterly *Policy Review*, the *Backgrounder*, and the *Heritage Lectures*, include studies on the uncertainty of global warming and the greenhouse effect.

The Intergovernmental Panel on Climate Change (IPCC)
C/O World Meteorological Organization, 7bis Avenue de la Paix, C.P. 2300, CH–1211 Geneva 2, Switzerland
+41-22-730-8208 • fax : +41-22-730-8025
e-mail : ipcc_sec@gateway.wmo.ch • website: www.ipcc.ch

Recognizing the problem of potential global climate change, the World Meteorological Organization and the United Nations Environment Programme established the IPCC in 1988. The IPCC's role is to assess the scientific, social, and economic information relevant for the understanding of the risk of human-induced climate change. The IPCC has published its Third Assessment Report, *Climate Change 2001*, in addition to special reports on global warming.

Pew Center on Global Climate Change
2101 Wilson Blvd., Suite 550, Arlington, VA 22201
(703) 516-4146 • fax: (703) 841-1422
website: www.pewclimate.org

The Pew Center on Global Climate Change is a nonprofit, nonpartisan, and independent organization dedicated to educating the public and key policy makers about the causes and potential consequences of global climate change. By releasing reports on environmental impacts, policy issues, and economics, the center works to encourage the domestic and international communities to reduce emissions of greenhouse gases. Its reports include *The Science of Climate Change*, *Human Health and Climate Change*, and *Sea-Level Rise and Global Climate Change*.

Rainforest Action Network (RAN)
221 Pine Street, Suite 500, San Francisco, CA 94104
(415) 398-4404 • fax: (415) 398-2732
e-mail: rainforest@ran.org • website: www.ran.org

RAN works to preserve the world's rain forests through activism addressing the logging and importation of tropical timber, cattle ranching in rain forests, and the rights of indigenous peoples. It also seeks to educate the public about the environmental effects of tropical hardwood logging. RAN publishes bimonthly *Action Alerts* and the biannual newsletter *World Rainforest Report*. Special RAN reports, such as *Drilling to the Ends of the Earth: The Ecological, Social, and Climate Imperative for Ending Petroleum Exploration*, offer a more in-depth look at causes of rain forest destruction and sustainable alternatives.

Rainforest Alliance
65 Bleecker St., New York, NY 10012
(888) MY-EARTH • fax: (212) 677-2187
e-mail: canopy@ra.org • website: www.rainforest-alliance.org

The Rainforest Alliance is an international nonprofit organization dedicated to the conservation of tropical forests for the benefit of the global community. Its mission is to develop and promote economically viable and socially desirable alternatives to the destruction of rain forests through education, research, and cooperative partnerships with businesses, governments, and local peoples. The alliance publishes the bimonthly newsletters the *Canopy* and *Eco-Exchange*.

Reason Foundation
3415 S. Sepulveda Blvd., Suite 400, Los Angeles, CA 90034
(310) 391-2245 • fax: (310) 391-4395
e-mail: keng@reason.org • website: www.reason.org/

Reason Foundation is a national public policy research organization that supports the rule of law, private property, and limited government. It believes that choice and competition will achieve the best outcomes in social and economic interactions. The foundation specializes in a variety of policy areas, including the environment, education, and privatization. It publishes the monthly magazine *Reason* and the studies *Evaluating the Kyoto Approach to Climate Change* and *Global Warming: The Greenhouse, White House, and Poorhouse Effects*.

Sierra Club
85 Second St., Second Floor, San Francisco, CA 94105
(415) 977-5500 • fax: (415) 977-5799
e-mail: information@sierraclub.org • website: www.sierraclub.org
The Sierra Club is a grassroots organization that promotes the protection and conservation of natural resources. In addition to numerous books and fact sheets, the Sierra Club publishes the bimonthly magazine *Sierra*, the *Environmental Currents* newsletter, and special reports including *Driving up the Heat: SUVs and Global Warming*.

Union of Concerned Scientists (UCS)
2 Brattle Square, Cambridge, MA 02238
(617) 547-5552 • fax: (617) 864-9405
e-mail: ucs@ucsusa.org • website: www.ucsusa.org
UCS works to advance responsible public policy in areas where science and technology play a vital role. Its programs focus on safe and renewable energy technologies, transportation reform, arms control, and sustainable agriculture. UCS publications include the quarterly magazine *Nucleus*, the quarterly newsletter *earthwise*, and the global warming reports *Greenhouse Crisis: The American Response* and *A Small Price to Pay: U.S. Action to Curb Global Warming Is Feasible and Affordable*.

World Resources Institute (WRI)
10 G Street NE, Suite 800, Washington, DC 20002
(202) 729-7600 • fax: (202) 729-7610
e-mail: lauralee@wri.org • website: www.wri.org
WRI provides information, ideas, and solutions to global environmental problems. Its mission is to encourage society to live in ways that protect earth's environment for current and future generations. The institute's program attempts to meet global challenges by using knowledge to catalyze public and private action. WRI publishes the reports *Climate, Biodiversity, and Forests: Issues and Opportunities Emerging from the Kyoto Protocol* and *Climate Protection Policies: Can We Afford to Delay?*

Worldwatch Institute
1776 Massachusetts Ave. NW, Washington, DC 20036
(202) 452-1999 • fax: (202) 296-7365
e-mail: worldwatch@worldwatch.org
website: www.worldwatch.org
The Worldwatch Institute is dedicated to fostering the evolution of an environmentally sustainable society in which human needs

are met in ways that do not threaten the health of the natural environment or the prospects of future generations. The institute conducts interdisciplinary and nonpartisan research on emerging global environmental issues such as climate change, the results of which are widely disseminated throughout the world. It publishes the annual *State of the World* anthology, the bimonthly magazine *World Watch*, and *Slowing Global Warming: A Worldwide Strategy* from the Worldwatch Paper Series.

Bibliography of Books

Nigel Arnell	*Global Warming, River Flows and Water Resources.* Chichester, England: Wiley, 1996.
Ronald Bailey, ed.	*Earth Report 2000: Revisiting the True State of the Planet.* New York: McGraw-Hill, 2000.
Roger Bate and Julian Morris	*Global Warming: Apocalypse or Hot Air?* London: Institute of Economic Affairs, 1994.
Melvin A. Benarde	*Global Warning . . . Global Warming.* New York: Wiley, 1992.
John J. Berger	*Beating the Heat: Why and How We Must Combat Global Warming.* Berkeley, CA: Berkeley Hills Books, 2000.
W. Bradnee Chambers	*Inter-Linkages: The Kyoto Protocol and the International Trade and Investment Regimes.* New York: University Press, 2001.
Alston Chase	*In a Dark Wood: The Fight over Forests and the Rising Tyranny of Ecology.* Boston: Houghton Mifflin, 1995.
Gale E. Christianson	*Greenhouse: The 200-Year Story of Global Warming.* New York: Walker and Company, 1999.
Jack Doyle	*Taken for a Ride: Detroit's Big Three and the Politics of Pollution.* New York: Four Walls Eight Windows, 2000.
Francis Drake	*Global Warming: The Science of Climate Change.* New York: Oxford University Press, 2000.
Christine A. Ennis and Nancy H. Marcus	*Biological Consequences of Global Climate Change.* Sausalito, CA: University Science Books, 1996.
Ross Gelbspan	*The Heat Is On: The Climate Crisis, the Cover-up, the Prescription.* Reading, MA: Perseus Books, 1998.
Ross Gelbspan	*The Heat Is On: The High Stakes Battle over Earth's Threatened Climate.* Reading, MA: Addison-Wesley, 1997.
Al Gore	*Earth in the Balance: Ecology and the Human Spirit.* Boston: Houghton Mifflin, 1992.
Michael Grubb, Christiaan Vrolijk and Duncan Brack	*The Kyoto Protocol: A Guide and Assessment.* Washington, DC: Brookings Institution, 1999.
Martin M. Halmann and Meyer Steinberg	*Greenhouse Gas Carbon Dioxide Mitigation: Science and Technology.* Boca Raton, FL: Lewis Publishers, 1999.

| John Horel and Jack Geisler | *Global Environmental Change: An Atmospheric Perspective*. New York: John Wiley & Sons, 1997. |

| John Houghton | *Global Warming: The Complete Briefing*. New York: Cambridge University Press, 1997. |

| Catrinus J. Jepma and Mohan Munasinghe | *Climate Change Policy: Facts, Issues, and Analyses*. New York: Cambridge University Press, 1998. |

| Jeremy Leggett | *The Carbon War: Global Warming at the End of the Oil Era*. London: Penguin, 2000. |

| Nick Mabey et al. | *Argument in the Greenhouse: The International Economics of Controlling Global Warming*. New York: Routledge, 1997. |

| Robert Mendelsohn and James E. Neumann, eds. | *The Impact of Climate Change on the United States Economy*. New York: Cambridge University Press, 1999. |

| Patrick J. Michaels and Robert C. Balling Jr. | *The Satanic Gases: Clearing the Air About Global Warming*. Washington, DC: Cato Institute, 2000. |

| Thomas Gale Moore | *Climate of Fear: Why We Shouldn't Worry About Global Warming*. Washington, DC: Cato Institute, 1998. |

| Michael L. Parsons | *Global Warming: The Truth Behind the Myth*. New York: Insight Books, 1995. |

| S. George Philander | *Is the Temperature Rising?: The Uncertain Science of Global Warming*. Princeton, NJ: Princeton University Press, 1998. |

| S. Fred Singer | *Global Climate Change: Human and Natural Influences*. New York: Paragon House, 1989. |

| S. Fred Singer | *Hot Talk Cold Science: Global Warming's Unfinished Debate*. Oakland, CA: Independent Institute, 1997. |

| P.C. Sinha, ed. | *Global Warming*. New Delhi, India: Anmol Publications, 1998. |

| P.C. Sinha, ed. | *Sea-Level Rise*. New Delhi, India: Anmol Publications, 1998. |

| Mark C. Trexler and Christine Haugen | *Keeping It Green: Tropical Forestry Opportunities for Mitigating Climate Change*. Washington, DC: World Resources Institute, 1995. |

| Karl K. Turekian | *Global Environmental Change: Past, Present, and Future*. Upper Saddle River, NJ: Prentice Hall, 1996. |

| Sylvan H. Wittwer | *Food, Climate, and Carbon Dioxide: The Global Environment and World Food Production*. Boca Raton, FL: Lewis, 1995. |

Index

224